INTERNATIONAL PROJECT ANALYSIS
AND FINANCING

International Project Analysis and Financing

GERALD POLLIO

Ann Arbor

THE UNIVERSITY OF MICHIGAN PRESS

Copyright © by Gerald Pollio 1999
Published in the United States of America by
The University of Michigan Press

2002 2001 2000 1999   4 3 2 1

Library of Congress Cataloging-in-Publication Data applied for
ISBN: 0–472–11095–0 (hardback)

Printed in Great Britain

For Rosina

Contents

List of Figures, Charts and Tables

Preface

This book grew out of lectures on project analysis and financing given by the author in classrooms, workshops and seminars organised by universities, financial institutions and training consultancies in Europe, the Middle East and the Far East. The earliest versions of these lectures were concerned overwhelmingly with natural resource projects, the industrial sector the author is most familiar with. Over time, the scope of the analysis and the case studies used to illustrate these principles have expanded in terms of both industrial and geographic coverage. Still, the book retains to a very considerable extent its original energy and mineral orientation. Nor is this inadvertent: natural resource sectors remain major users of limited recourse debt, while the fallout from the recent Asian financial crisis could easily result in the recovery of their former pre-eminent market position.

Numerous articles and books have been written over the past decade or so covering virtually every facet of the project finance market in varying degrees of comprehensiveness. A glance at the references given at the back of this book may surprise some readers, indicating as it does how widespread the literature on this topic is. Accordingly, some justification may be needed to explain the appearance of yet another study on a subject that looks so well catered for. What differentiates the present study from most others is the conviction that project analysis and financing are two sides of the same coin. This may sound obvious, but in practice authors who write about project financing normally assume readers possess an understanding of the principles of investment analysis and *vice versa*. This book makes no such assumption. Indeed, the title is intended to convey the author's firm belief that the two subjects are of equal significance. Having said that, it should be emphasised that readers are not expected to have a background in finance, only an interest in the subject. All of the principles needed to make sense of this book are developed as an integral part of the analysis.

In the course of writing this text, I have acquired numerous intellectual debts. As already indicated, the background to this study is a series of lectures given at various places over the past several years. Participants, I am pleased to say, responded positively to the lectures. I would, however, like to express my appreciation to all those individuals who by their questions or suggestions helped to clarify and ultimately improve the content and presentation of the material contained in this book. I was indeed fortunate in being able to incorporate into the present volume the comments of so many individuals made over so long a period of time. One individual, however, should be singled out by name. Professor Roger Wooten, Dean of the School of Engineering, City University, read and commented upon an earlier version of this study. Finally, an earlier version of Chapter 5 was published in *Energy Policy*.

Marlow, Buckinghamshire   GERALD POLLIO
July 1998

Acknowledgements

The author and publisher wish to thank the following for permission to use copyright material:

Jonathan Drew (Hong Kong Shanghai Bank) for Table 1.1 from Recent Bond Ratings and Asian Sovereign Ratings, and Table 1.2 from Project Bond Ratings and Structures, 1996/97 and Table 1.3, Panel A.

HSBC (Hong Kong Shanghai Banking Corporation) for Table 1.4 from *Capital Markets Funding for Limited Recourse Debt: A Guide to the Rule 144A Market* (1998).

John Wiley and Sons, Inc. for material from Finnerty, *Project Financing* (1996) and Figure 2.1 from Bernstein, *Against the Gods* (1996) The Blessings of Diversification (p. 234).

Elsevier Science for Tables 1.6, 1.7 and 1.8 from Lang (ed.), *A Comparison of Project Finance in Asia and the West* (1997) and for material in Chapter 4 from *Energy Policy*, 'Project Finance and International Energy Development', Pollio, Vol. 26.

Oxford University Press for Table 2.3 from Robert Barro, *Review of Financial Studies* (1990), 'The Stock Market and Investment'.

Prentice Hall for Table 2.4 from Emery and Finnerty, *Corporate Financial Management* (1997).

Stern, Stewart Management Services for Tables 3.3 and 3.4 from Siegel, Smith and Paddock, 'Valuing Offshore Properties with Options Pricing Models', *Midland Journal of Corporate Finance* 1987; Table 3.7A from *Journal of Applied Corporate Finance*, **9**, 1996, 'Incorporating Country Risk in the Valuation of Offshore Projects', Lessard; Table 3.7B from 'A Practical Approach to Calculating Costs of Equity for Investments in Emerging Markets', Godfrey and Espinosa; Table 6.4 from *Journal of Applied Corporate Finance*, **5**, 1992, 'Project Valuation Under Uncertainty: When Does DCF Fail', Kulatilaka and Marcus.

International Monetary Fund for Tables 3.5 and 3.6 from Staff Papers, 'The Economic Content of Indicators of Developing Country Creditworthiness', Haque, Kumar, Marx and Mathieson, 1996.

Standard and Poor for Tables 7.1, 7.3 and 7.4 from *Global Project Finance*, September 1997.

Confederation of British Industries for Table 2.2 from *Realistic Returns: How do Manufacturers Assess New Investment?*

Every effort has been made to trace all the copyright holders but if any have been inadvertently overlooked the publishers will be pleased to make the necessary arrangements at the first opportunity.

1 An Overview of Project Analysis and Financing

This study is about project analysis and financing. Project analysis is concerned with identifying and assessing the value-enhancing potential of individual investment opportunities. Project financing in its broadest sense encompasses all sources of funds used to finance project investments. It thus includes equity and debt, straight bank loans and bonds as well as non- or limited recourse financial structures, again either bonds or bank loans. It is the latter, narrower class of liabilities that forms the core of the present study.

The interconnections that exist between project analysis and financing, although obvious, are all too often ignored. Indeed, the two are so closely linked it could be argued with only slight exaggeration that project finance is largely incomprehensible without a thorough understanding of investment analysis. The recent financial crisis in east Asia provides a powerful example of how close the link really is. Ironically, it does so by showing that the origins and progress of the crisis can be explained largely or entirely in terms of the commercial loan market's failure to integrate the two concepts. This is manifest in the abandonment of traditional credit principles, above all, the imputation of greater than warranted value to (mostly implicit) official guarantees. Commercial lenders, like the industrial investors whose projects they were financing, gave full faith and credit to such guarantees.

To appreciate the importance of such undertakings, we need first to introduce the concept of maximum loan value. Maximum loan value is defined as $L = C/\varphi$, where L is the maximum amount of debt a lender would ascribe to a project given its cash generating potential (C), and φ, the desired coverage ratio, is defined as C/L. The desired coverage ratio is a parameter, which tends to be invariant across individual lending institutions. Now, according to Krugman (1998) Asian projects were typically evaluated not on the basis of expected returns – that is, those suggested by traditional investment or credit analysis – but rather in accordance with a given project's 'Pangloss' value, defined as returns obtainable under the best of all possible circumstances.[1]

Three key implications follow from lending and investing decisions based upon Pangloss valuations. First of all, the level of regional investment will

1

be far higher than could be justified economically. Also, the level of lending commitments will be similarly inflated. Indeed, the difference in the scale of lending commitments is given by $(C^* - C)/\varphi$, where C^* and C are the cash flow potentials based upon, respectively, the project's Pangloss and expected values. The latter are derived from more conservative investment criteria, for example, those suggested by traditional financial analytical methods.[2] And finally, assets that are in limited supply, real estate for example, will be bid up to well above fair market values.

The process (or 'bubble') can continue, according to Krugman, only so long as government guarantees are credible. Credibility, like bankruptcy to which the concept relates, is an inverse function of the level of outstanding official guarantees. At the point where the authorities are obliged to cover actual losses, credibility is destroyed. Rising bailout costs diminish the authorities' ability to provide for future financial rescues; asset values, accordingly, collapse. The scale of potential losses is vast, some indication of which is given by the magnitude of the financial stabilisation programmes currently being negotiated between the International Monetary Fund and some of the region's largest economies.[3]

The above example underscores the close relationship that exists between real and financial decisions, and how the two can interact to cause financial panics or worse. The analysis also provides a rough quantitative estimate of the costs to banks connected with departures from the sound financial principles upon which regional lending decisions should have been based. The reverberations of the crisis are still being felt in the project finance market. For the commercial loan market, a recent tabulation indicates that in 1997 loans aggregating $151.1 billion were arranged for 384 projects; the corresponding figures for 1996 were $223.9 billion and 894 projects, respectively (Project Finance, 1998).

As to the project bond market, Table 1.1 summarises some of the more salient statistics. Note, in particular, the sharp deterioration in country ratings: 63 per cent of new Asian issues were rated BB or lower at year end 1997; one year earlier, the comparable figure was only 26 per cent. Ratings spreads among the region's main borrowers have also widened considerably. In mid-year 1997 – that is, just prior to the devaluation of the Thai bhat which triggered the financial crisis – the maximum regional spread over US Treasuries amounted to around 100 basis points; by the first quarter of 1998, the gap had widened to more than 500 basis points (Drew, 1998). As the last panel of the table shows, market assessment of the creditworthiness of individual east Asian borrowers now ranges from 'strong' for Singapore to 'predominately speculative' for Indonesia.

There is no doubt the project finance market will eventually recover from recent reverses. It is less clear whether financial markets will resist the temptation to overreact to recent events. If they do, the crisis will have produced positive benefit.[4]

Table 1.1   Recent bond rating changes, new Asian issues and project bonds in US capital markets

Rating	1 December 1997	1 January 1998
	1997 issues from Asian credits	
AAA	9%	
AA	14%	
A	30%	22%
BBB	21%	15%
BB	22%	19%
B	4%	44%

Asian sovereign ratings: Standard & Poor's *long-term foreign currency debt*

Country	July 1996	September 1998
China	BBB	BBB+
Hong Kong	A	A
Indonesia	BBB	CCC+
Korea	AA−	BB+
Malaysia	A+	BBB+
Philippines	BB	BB+
Singapore	AAA	AAA
Taiwan	AA+	AA+
Thailand	A	BBB−

Source: Drew (1998).

LIMITED RECOURSE FINANCE

Numerous advantages are claimed for limited recourse debt over rival debt vehicles, primarily the ability to generate benefits for project sponsors above and beyond those attainable within the framework of more traditional financial structures. These alleged benefits include such things as maximising potential tax advantages, addressing agency problems that might arise in connection with multi-party sponsorship, protecting corporate existence from the risk of major project failure, and so forth. No matter how plausible such arguments sound, the available empirical evidence suggests, as we shall see later, that the importance of these benefits individually or collectively is still very much an open question.

An alternative and perhaps more compelling possibility is that project finance, like debt structures generally, is capable of creating strategic advantage for borrowers. In other words, project finance can be used to enhance prospective equity returns, an apparently counter-intuitive proposition until we recall that debt itself is one of the key determinants of required investment returns. Debt assigns more project risk to equity-holders, who accordingly demand higher compensatory returns. To the extent lenders are aware – as they must be – of the potential strategic benefits limited recourse financing can create, we might expect creditor interests to be safeguarded in ways fundamentally different from those that apply to secured borrowings. Indeed, it is precisely the resulting complex web of contractual and other legal arrangements more or less unique to this type of lending that differentiates limited recourse from more traditional forms of borrowing.

In subsequent chapters these themes and the issues that follow directly from them will be developed and analysed in greater depth, enabling the reader to decide which of the prevailing interpretations provides the most convincing explanation as to why sponsors choose project finance over competing borrowing options. For present purposes, the key point to remember is that the distinguishing feature of project finance debt is that it is limited recourse; hence the use of this descriptor as an alternative to the more common but less accurate 'project finance' appellation.

Limited recourse means that creditors (commercial lenders or other providers of fixed or floating rate debt), who typically provide the lion's share of the funds to construct the project, look mainly or only to the investment's potential cash flow to repay the loan.[5] In other words, funds' providers have no or only limited recourse to the investment's sponsors once certain predetermined technical, operational, economic and financial tests imposed by the creditors on the project have been met. Prior to completion, project loans will normally be fully guaranteed by the investment's sponsors.

The importance of completion cannot be emphasised enough: this feature effectively transforms what under other circumstances would be equity risk to operating risk. Operating risk can be defined for present purposes as the possibility the project will be unable to produce output in the quantity or quality or at a cost that ensures sufficient cash flow generation to repay the original indebtedness. Creditor exposure, in other words, is limited to that phase of the investment where the project is up and running and all predetermined completion requirements have been satisfied. Project risks are, of course, fundamentally different in the pre- and post-completion phases. Investors bear all project risks during construction, when the project is consuming cash without generating any revenues – the true focus of equity risk.

This distinction is important for another reason as well: it goes a long way toward explaining the modest margins applying to limited recourse loans compared with more traditional balance sheet financings. If creditors did in fact bear equity risks, prevailing spreads would obviously provide lenders with totally inadequate compensation. Once we recognise the actual risks to which they relate, market premiums appear far more appropriate.

Finally, many project finance studies assume that any financing that depends mainly or exclusively on the cash flow from a self-contained investment is equivalent to a limited recourse financing in the sense defined above. Viewed this way it is possible to argue, as many authors have, that such structures constitute a very ancient form of lending. To prove the point they cite numerous historic examples of the use of project finance to fund domestic and international investments including, significantly, natural resource and infrastructure projects as well as a wide range of maritime transactions. Buckley (1996b), for example, claims that loan structures approximating to what today would be described as limited recourse loans were used to finance construction of the Suez Canal. An even earlier example according to Finnerty (1996) is that the Frescobaldi, an Italian banking house, provided project finance loans in the late thirteenth century enabling the English government to develop silver mines in Devon. Repayment took the form of giving lenders the right to extract as much unrefined ore as could be mined in a year. The Crown provided no guarantees of any kind in respect of the quantity or quality of ore reserves, suggesting to Finnerty that lenders were prepared to accept reserve risk much as they would in a modern production loan. Lastly, McKechnie (1990) has argued that limited recourse transactions can be identified in ancient Athens, where *trapeza* financed grain and other imports but were entitled to recover capital and interest only if the voyage were completed successfully.

The significance of these examples as early instances of project finance lending is compromised once we recognise a confusion common to all three studies, namely, the failure to draw a clear distinction between a single purpose company formed to realise a given project, and a company whose cash flow is generated by a single project.[6] Limited recourse lending applies only in the first instance, with the additional proviso that lenders usually do not bear any equity risks within the framework of the financing. On this basis it is extremely doubtful whether the transaction discussed by Finnerty has any relevance at all to project financing in the narrow sense used above. Since usury was prohibited in the thirteenth century, it could of course be argued that the structure of the financing, with its prominent emphasis on equity risk, was simply a disguised attempt at circumventing church law, possibly an early example of financial engineering to address a key project risk! On the other hand, there is no compelling reason not to accept the transaction at face value: in return for providing mine finance, lenders acquired upfront a potentially large, but uncertain, share of project output.

McKechnie's example, too, seems more convincing as an equity investment than a limited recourse loan. Certainly, the widespread syndication of such transactions – suggesting, incidentally, that Athenian *trepazitai* were well aware of the benefits of portfolio diversification – and the 20–30 per cent returns earned on individual maritime loans points in that direction. Even if we concede, as current research indicates, that banking practices in ancient Athens were far more advanced than is commonly supposed, what data there are provide no support whatever for the hypothesis that project finance loans originated in Greece: 'although no law or regulation prohibited "nonrecourse" financing, not a single example is known of such lending in Athens' (Cohen, 1992). Nor is Buckley's example convincing: once we take into account the significant concessions and support provided by the Egyptian government to advance the project, it is more reasonable to view the financing as a sovereign credit than a project finance loan.

The simple fact that most ventures historically were developed on a project by project basis, the observation that motivates these and most other putative examples of early project finance loans, provides no more compelling evidence of the existence of limited recourse lending than does the prevalence of traditional balance sheet financing prove the opposite. Limited recourse financing is a clearly delineated loan structure, supported by well-defined contractual arrangements. Whether historic transactions conform to what we would now describe as limited recourse financing depends crucially upon being able to establish the existence of *all* the requirements that define this type of lending.

SOURCES OF LIMITED RECOURSE FINANCE

Sponsors now have the flexibility to source funds to construct projects in a number of global markets. Historically, the project finance market consisted almost exclusively of loans provided by the international banking system. Project loans were underwritten by leading multinational banks; in many instances these banks were responsible for both arranging and providing the project loans. This was not always or even necessarily the case. In fact, the project finance loan market today is effectively partitioned into those institutions that are arrangers and those that are funds' providers. North American and west European banks typically fall into the former category with Japanese banks, although among the largest in the world, overwhelmingly providers of funds. Only in a very few instances do individual banks straddle both categories.

Investors have recently begun to tap international bond markets for project finance as an alternative to the more traditional commercial loan market. Some commentators assert that the resulting increase in competition will compress margins at a time when existing project portfolios are under pressure from the east Asian financial crisis. Indeed, the fact that losses on international bonds – even by sovereign borrowers – have historically been small compared to the experience of leading multinational banks highlights the superior position of the former (Pollio, 1992). This interpretation, however, is unduly pessimistic, as it fails to take account of the complementarities that exist between the two markets. These complementarities could just as easily imply an increase in project debt capacity, thus supporting a higher level of investment activity than would otherwise be possible (Middlemann, 1996).[7]

The significance of this latter point is highlighted by the structure of recent major gas financings in Qatar, a small Arabian peninsula oil and gas producing country with no past history of having raised significant sums in international financial markets. By way of background we might note that, in 1994, the downstream phase of Qatargas, one of two major liquefied natural gas (LNG) projects then under construction, was financed by a consortium of Japanese banks. The upstream phase of the project, involving development of the field from which both Qatargas and its companion Ras Laffan project would source their respective feedgas supplies, was hived off from the downstream project and financed instead as a condensate project by a syndicate led by French banks. The lending syndicates formed to finance each project phase were hardly accidental. A Japanese utility company was the sole purchaser of downstream output, a Japanese contractor was chosen to build the liquefaction facility, and the project enjoyed strong official support. For the upstream financing it is significant that Total, an integrated French energy company, had an equity stake in both the production and downstream phases of the project.

British Petroleum was one of the project's original foreign joint venture partners. Owing to acute financial difficulties experienced in the early 1990s, British Petroleum decided to drop out of the venture. Mobil replaced BP and was equally keen to develop a second LNG project. Mobil's interest in the concurrent development of two major LNG projects in Qatar can only be explained by growing reserve difficulties in Indonesia, where the bulk of its LNG investments up to that time were concentrated. We can only assume that although alternative Indonesian options were available, cost and other project-specific considerations led Mobil to favour Qatar.

Development of two LNG projects in Qatar was likely to stretch market capacity. In other words, it is arguable whether international banks would have been willing to increase their existing Qatari exposure by the full amount of new project debt that sponsors were seeking to raise. Capacity pressures were further compounded by the fact the Emirate was at the same time looking to finance a number of gas-based manufacturing projects, as well as to upgrade and expand its local refinery. The health of the economy was also a matter of concern: the Qatari government was running huge budget deficits, owing mainly to the need to finance the national oil company's majority stake in local energy and petrochemical projects. Moreover, the government was financing its deficits with bank overdrafts thus creating acute local liquidity problems.

It was at this point the possibility of including an international bond tranche in the financing of Mobil's Ras Laffan project was first investigated. As things turned out, the market was extremely receptive, the issue over-subscribed and the pricing considerably more favourable than envisaged when the idea was first mooted.[8] Strong international interest in the issue was hardly surprising, as the bonds provided investors with a more or less pure commodity (natural gas) exposure linked to the growth of the east Asian economies. What was initially perceived to be the issue's primary appeal turned, ironically, into the investment's principal weakness: the east Asian financial crisis, with its attendant negative impact on medium- and longer-term growth prospects, and hence future gas demand, recently resulted in the issue being placed on the watch list.[9]

Other project issues have appeared subsequent to Ras Laffan; even so international bond markets provide only a very modest share of total limited recourse finance. Moreover, the market tends to be concentrated over-whelmingly in the United States. As of the third quarter of 1997, the number of outstanding international project bonds amounted to less than 100 issues with a combined market value of $35 billion; roughly two-thirds were arranged on behalf of US borrowers (HSBC, 1997). Nearly one half of outstanding project issues have been rated by Standard & Poor as BB or higher, although as the data given in Table 1.2 show, there has been an increase in the market's appetite for riskier issues; note in this connection that 86 per cent of outstanding issues were rated BBB or higher in 1996. There is also evidence of a concomitant willingness to favour single payment over amortising bonds as the near doubling between 1996 and 1997 of bullet issues attests.

Growth prospects for international project bonds are generally viewed more favourably than are more traditional project loans. This assessment

Table 1.2   Project bond ratings and structures, 1996 and 1997

Rating	Definition	1996	1997
A	Strong	59%	1%
BBB	Adequate	27%	45%
BB	Predominately speculative, to varying degrees	18%	50%
B	Predominately speculative, to varying degrees	0%	4%
Structure			
Bullet		23%	42%
Amortising		77%	58%

Source: Scarlett (1990) and Drew (1998).

appears to be based less on the market's putative flexibility and cost advantages, than on the prospective continued rapid growth of global infrastructure projects coupled with the expectation that the majority of such investments will continue to be funded on a project finance basis. In other words, the market for limited recourse debt is expected to continue to expand rapidly, while the demand for international bonds is projected to grow even faster.[10] The principal assumption underpinning this view appears much shakier now than it did even as recently as a year ago. The market for infrastructure projects should be much less robust going forward than assumed in pre-crisis forecasts. This means that, if recent projections are to materialise, the project bond market will have to gain at the expense of the more traditional commercial loan market, an outcome, incidentally, that leading international banks might not view all that negatively.

Whether this in fact happens depends upon the significance of a relatively small number of variables that influence the choice of financing vehicle. Panel B of Table 1.3 compares the commercial loan market, the private placement market, and the quasi-public Rule 144A market in terms of six principal characteristics.[11] Each market has its own positive features, but taken together it is easy to see why many commentators favour the continued rapid growth of the project bond market, above all, the quasi-public market in the US. This latter market in particular exhibits a far greater degree of flexibility than do either the commercial loan or private placement markets: it is at least as liquid as the public market, maturities tend to be longer – up to 30 years compared with a maximum of 15 years in the loan market – arranging times are shorter and financial covenants less restrictive.

Table 1.4 provides additional information on the characteristics of leading international fixed income markets. Each market is unique thus providing international borrowers with a very broad menu of funding options. Note, in particular, how the investor base varies across different geographic regions; in each market issuers are also able to raise finance not only in regional currencies, but also in US dollars or some combination of the two. The US market is far and away the most sophisticated and liquid of leading international fixed income markets. For example, the US market is capable of offering borrowers maturities up to 100 years as against a maximum of 30 years elsewhere. The size of the US market, as measured by issues outstanding, amounts to roughly $15 trillion, the overwhelming majority of which consists of government debt. The non-government sector, while accounting for only about 11 per cent of the market, still aggregates in absolute amount to more than $4 trillion.

International borrowers now rely mainly on the quasi-public segment of the market, which in 1997 accounted for 45 per cent of international issues, nearly twice the private placement share and 14 percentage points higher than public issues. The Rule 144A market is dominated by insurance

Table 1.3   Comparison of alternative funding sources

Panel A: General characteristics

Traditional funding sources

Syndicated loan market	*Export credit agency market*
● Direct involvement in the process with substantial control – 'hands-on' management of independent consultants.	● Direct involvement in the project with substantial control.
● Ability to handle complex funding arrangements and perform own credit analysis; no rating required.	● Ability to handle complex funding arrangements, possibly involving an adviser, typically require commercial banks to share risks; no rating required.
● Investment parameters are typically limited to floating rate funding, limited maturities and repayment schedules.	● Long-term fixed rate funding available for sourcing of equipment.
	● Certain principles may be inflexible.

Capital markets funding sources

Unrated private placement market	*Rated 144A transactions*
● Some investor involvement in the project but with limited control.	● Relatively little or no investor involvement in the project with objective/standardised requirements.
● Ability to handle complex credit situations or transactions where sophisticated structuring improves project economics.	● With increased understanding and market demand, capital market investors participate in projects with construction risk and emerging market credit issues.
● Buy and hold investors who do not require ratings or secondary market liquidity.	● An increasing universe of investors as a result of obtaining a rating and increased liquidity associated with a Rule 144A transaction

companies and fund managers, accounting for 40 per cent and 35 per cent of the market, respectively; state and local pension funds, at 12 per cent, constitute the third largest market segment.

Each market segment tends to be parochial, focusing on specific borrower characteristics or issues. Insurance companies, for example, tend to prefer A rated or weaker credits while state and local pension funds concentrate only on issues rated A to AAA; fund managers, by contrast, appear to be the least credit sensitive, purchasing issues across the entire ratings spectrum. Finally, insurance companies and pension funds favour longer-term issues in contrast to bank trust departments, which tend to focus on short to intermediate maturities (HSBC, 1997).

Table 1.3 (continued)

Panel B: Specific characteristics of alternative project finance markets		
Characteristic	Commercial bank market	Private placement market	Rule 144A quasi-public market
Maturity	Up to 15 years.	Up to 20 years.	Up to 30 years.
Interest rate	Floating rate. Interest rate risk can be eliminated via interest rate swap.	Fixed or floating rate.	Fixed or floating rate.
Prepayments	Permitted, subject to unwinding the interest rate swap, if any.	Requires compensation for any lost future interest income.	Normally permits greater refunding flexibility than the commercial bank or private placement market.
Covenants	Comprehensive set of financial covenants.	Comprehensive set of financial covenants.	Usually less restrictive financial covenants than applicable in commercial bank or private placement markets.
Time to arrange	15–25 weeks.	10–20 weeks.	10–15 weeks.
Reporting requirements	None.	Generally requires an NAIC-2 rating.	Generally requires investment grade ratings from at least two major rating agencies.

Source: Drew (1998) and Finnerty (1996).

RECENT TRENDS IN PROJECT FINANCE LENDING

We conclude this overview with a summary of the historic and current structure of the limited recourse loan market. Table 1.6 presents data on a large sample compiled by Kleimeier and Megginson (1997) of project loans arranged over the period 1977 to 1992. The petroleum industry and related sectors accounted for roughly 30 per cent of all internationally syndicated project loans over the 15 year period; adding in other mineral industries, the proportion rises by an additional 15 percentage points. Power projects, which may, but need not, tie in with primary energy development – that is, power projects can involve the concurrent development of new or existing

Table 1.4　　International fixed income markets

Market	Principal characteristics	Investor base
US market	Deep floating and fixed rate market. Maturity parameters: 2–100 years; primary currencies: US dollar.	Pension funds, fund managers, insurance companies, banks and other domestic and foreign investors.
Euromarket	Deep floating and fixed rate market. Maturity parameters: 2–10 years; Primary currencies: major European, US dollars.	European banks, central banks, money managers, fund managers and insurance companies.
Asian market	Deep floating and fixed rate market. Maturity parameters: 2–30 years; Primary currencies: major Asian, major European, US dollars.	Trading companies, Asian banks, insurance companies, money managers, fund managers and central banks.

Source: HSBC (1997).

hydrocarbon resources – accounted for an additional 25 per cent of the market. Given this structure it is hardly remarkable that mineral producing regions accounted for the bulk of international project loans, up to 55 per cent according to the second panel of the table. Finally, the mean (median) project loan arranged over the period amounted to $426 million ($200 million); the largest single project loan was $7 billion, while the smallest was $2 million.

Table 1.5 provides more recent (non-comprehensive) data on the composition of the project loan market. The most striking features of the table are the relative decline in the share of loans arranged on behalf of traditional borrowing sectors, and the concomitant increase in the proportion of total loans accounted for by local infrastructure projects. In Europe, for example, natural resource and allied sectors accounted for just under one-fifth of the market in 1995, with the petroleum industry representing around half of the sectoral total. Power projects, by contrast, now account for almost two-fifths of total project loans; adding in telecommunications and other infrastructure loans, the combined share approaches 70 per cent. Elsewhere, the pattern is more or less in keeping with historic trends: three-fifths of identifiable emerging market financings (signed agreements plus in finance) over the past two years were in traditional sectors, with petroleum and petrochemical projects accounting for roughly 60 per cent of that total. Loans arranged on behalf of the telecommunications and power industries aggregated $35 billion, a figure only marginally below the petroleum sector total.[12]

The United States remains far and away the largest single borrower in the project finance market, with transactions aggregating $47 billion in 1996, that is, some 2.5 times the amount arranged for Hong Kong, the second

Table 1.5 International project loans, 1996

Panel A: Project loans by region		
Region	*Project count*	*Amount (US$m)*
Asia–Pacific	353	76,263
North America	126	51,139
Western Europe	92	34,285
Latin America and Caribbean	101	21,473
Eastern Europe and FSU	120	15,640
Middle East	23	12,225
South Asia	44	9,854
Africa	36	3,018

Panel B: Project finance loans by sector (billions of dollars)		
Sector	*Europe*	*Emerging markets*
Power	8.7	17.2
Telecommunications	5.5	17.9
Oil and gas	2.4	38.1
Infrastructure	2.0	
Industrial	1.5	
Mining	1.1	
Petrochemicals	0.9	17.6
Leisure	0.7	
Others	0.4	

Source: Panel A: *Project and Trade Finance*, March 1997. Panel B: Europe: Taylor (1996); Emerging markets: *Project and Trade Finance*, various issues.

Note: Figures in Panel B, Europe are for 1995, and for emerging markets 1995–96. The emerging markets data include both signed deals and loans in negotiation.

largest borrower. Among other emerging market economies only Indonesia ranked among the top five borrowers, although significant additional sums were raised for countries in south or east Asia. Both Thailand and China, for example, obtained up to $9 billion in 1996 in the project finance market. On average project loan values, Qatar at $1.6 billion tops the sample. These loans were arranged on behalf of Qatar's two major LNG projects and actually understate the total since the financing for one of these (Ras Laffan) includes a $1.2 billion international bond tranche.

Given the importance the east Asian financial crisis has had and will continue to have on limited recourse markets it might be useful to summarise what is known of the historic structure of regional transactions. To this end, Kleimeier and Megginson (1997) recently reclassified their original sample into Asian and non-Asian loans, with the results shown in the following tables. Table 1.6 presents data on Asian and non-Asian loans based upon two classifications, whether the data series includes or excludes loans under active negotiation; a somewhat larger fraction of Asian loans included in the Kleimeier and Megginson sample were under negotiation, a fact having no major bearing on the reported results. Asian borrowers account for just under 40 per cent of the Kleimeier and Megginson sample. Australia emerges as the leading regional borrower accounting for between 23–29 per cent of the regional total, depending upon which measure is used. Among emerging east Asian borrowers, Indonesia, Malaysia, Thailand and the Philippines stand out.

Natural resource and allied projects – broadly defined to include mining, oil and gas, refining, chemicals and primary metals – account for a slightly larger share of Asian than non-Asian projects, although again the results tend to be dominated by Australia, which accounts for seven of the eleven mining projects included in Table 1.7. Among traditional market sectors, Asian projects dominated the non-Asian sample in only two traditional sectors, namely, mining (11 projects) and chemical manufacture (10 projects); transportation was the only other sector where Asian projects predominated, with 12 projects as against two for the non-Asian sample.

On the other hand, non-Asian borrowers dominated the Asian sample in only two sectors, namely, oil and gas extraction and power. However, it should be noted that, in the latter sector, nine of the reported project loans are in the United Kingdom; excluding the UK, the sectors would be much more closely aligned. It is doubtful whether the same conclusion applies in respect of oil and gas extraction, where North America and the North Sea almost certainly dominate the total. The first panel of Table 1.8 compares the Asian and non-Asian sample in terms of a number of major loan characteristics; panel B quantifies whether the observed differences are statistically significant or are due to chance.

The impression gained from panel A is that Asian loans carried slightly larger spreads, had slightly shorter tenors and substantially smaller loan values. More surprising is that non-Asian loans had a somewhat higher ratio of guaranteed to non-guaranteed borrowers and a smaller number of project sponsors. Once these differences are subjected to quantitative analysis, the only differences that show up as being statistically significant are loan size, number of sponsors and the year in which the loan was booked. Even excluding the Eurotunnel financing, the first finding looks robust in that the non-Asian sample mean declines to only $410.5 million, still nearly 50 per cent higher than the mean sized Asian loan.

Among other points worth noting, the largest Asian loan was significantly smaller than its non-Asian counterpart, a finding that appears to hold even if the Eurotunnel loan is again excluded, as were loan tenors, with a maximum of 18 years for the non-Asian sample, some 3.5 years longer than the maximum Asian tenor. In contrast, the maximum spread on an

Table 1.6   Project loans by country or region, 1977–96

Location	Number of project finance loans booked plus in negotiation	Number of project finance loans booked only
Asia		
Australia	42	39
Indonesia	26	20
Malaysia	20	19
Thailand	18	12
Philippines	14	7
Turkey	12	3
China	10	7
India	8	5
Pakistan	8	4
Hong Kong	5	0
New Zealand	5	5
Singapore	3	3
Bangladesh	1	1
Laos	1	0
South Korea	1	1
Uzbekistan	1	1
Asia total	181	135
Non-Asia		
United States	57	53
United Kingdom	37	33
North Sea	29	29
Russia (includes FSU)	17	13
Canada	15	14
Germany	8	7
Colombia	5	3
Hungary	5	3
Nigeria	5	4
Other non-Asian countries	65	50
Non-Asian total	243	209
Unknown country	43	42
Total, all countries	467	386

Source: Kleimeier and Megginson (1997).

Table 1.7   Industrial classification of the Kleimeier–Megginson project finance loan sample

		Number of project finance loans	
SIC code	Industry	Asia	Outside Asia
10,12,14	Mining	11[a]	7
13	Oil and gas extraction	8	15
15	Building construction	0	3
25	Furniture manufacture	1	0
26	Paper manufacture	3	0
28	Chemical manufacture	10	0
29	Petroleum refining	3	4
33	Primary metals	2	2
35	Ind. and comm. mach.	1	0
39	Misc. manufactures	1	1
41, 44, 45	Transportation	12	2
46	Pipelines	3	5
49	Power generation	2	13[b]
70	Hotels	2	1
79	Recreation services	2	1
80	Health services	0	1
83	Water treatment	0	1
89	Misc. services	0	1
–	Unknown industry	1	1
	Total	62	58

Source: Kleimeier and Megginson (1997).

Notes:
a   Seven of the 11 loans shown for the category are loans to Australian projects.
b   Nine of the 13 loans indicated are loans to UK projects.

Asian loan was 17 basis points lower than the highest observed non-Asian spread, while the maximum number of sponsors for a non-Asian project, at eight, was twice the maximum Asian figure.

All in all, the results are not far removed from what might have been expected *a priori*: the Asian region was historically a major user of project finance. On the other hand, transactions on average were smaller than those arranged for non-Asian countries but otherwise exhibit characteristics similar to those reported for other regions. It is unfortunate that the authors did not restrict the comparative analysis to emerging markets only, surely the more relevant comparison. Had they done so, we may be confident the observed differences would have been far more telling, highlighting why the crisis has had such a significant impact on market sentiment and on the scale of new financings.

Table 1.8   Asian and non-Asian project loans, 1977–92

Panel A: Characteristics of project finance loans in Asia and the West

Characteristic Mean (range)	Asian borrowers	Non-Asian borrowers
Spread (in bp)	107.3 (12.5–282.5)	101.6 (12.5–300.0)
Year	1989 (1981–92)	1986.9 (1975–92)
Tenor (years to maturity)	10 (3–14.5)	10.56 (1–18)
Sponsors (no. of companies)	2 (0–4)	3 (1–8)
Size (US$ millions)	274.5 (9.0–2 360.0)	523.2[1] (12.0–6 720.0[2])
Guarantee (1=yes, 0=no)	0.517 (0–1)	0.435 (0–1)

Panel B: Tests for differences between Asian and non-Asian project finance loans

Variable	Sample size Asia	Non-Asia	Wilcoxon rank sum test Z-value	Median test Z-value
Spread	62	58	−1.002	−1.455
Year	62	58	−2.626[a]	−2.022[b]
Tenor	57	52	0.256	0.223
Sponsor	22	15	1.794[c]	2.212[b]
Size	59	56	2.946[a]	1.948[b]
Guarantee	29	23	−0.575	−0.585

Source: Kleimeier and Megginson (1997).

Notes:
1 Excluding the Eurotunnel loan, the non-Asian mean would be $410.5 million.
2 Excluding the Eurotunnel loan, the non-Asian maximum would be $3 000 million.
3 a, b and c denote statistical significance at the 1, 5 and 10 per cent level, respectively.

PLAN OF THE BOOK

The following chapters explain in detail the concepts that were introduced in preceding sections, and which are needed to achieve a clearer grasp of the significance of the data just presented. Chapter 2 introduces the principles of capital budgeting, that is, the criteria used by project sponsors to implement investment opportunities. It goes without saying that projects that do not make sense as equity investments – those, in other words, unlikely to produce acceptable returns to sponsors – can never be of interest to project finance markets. Limited recourse, the defining characteristic of project finance debt, means that upon completion sponsors have the right to abandon the project in favour of the debt providers. If a project is unprofitable, investors will have a powerful incentive to exercise their option to default.

One very important concept, central in fact to a proper understanding of capital budgeting, is that of opportunity cost. Opportunity cost arises because economic resources are scarce, and hence valuable, and can be deployed in different uses. Not all assets have an opportunity cost; those that do not fall into the category of sunk costs, costs that are unrecoverable and hence of no further concern to investors. Seen this way, most investment projects represent prospective sunk costs. Once committed to a specific purpose, project assets can no longer be used elsewhere. Among the primary objectives of capital budgeting theory is the determination of equity returns required by sponsors to invest in a given project as opposed to other investment opportunities that may also be available.

The simplest criterion of all is that a given project should more than cover the opportunity cost of the capital committed to the investment; in modern finance theory this is known as the Net Present Value rule. If prospective returns just equalled those obtainable on the next best opportunity, investors would be indifferent as to which they chose. If, by contrast, expected returns fell short of those capable of being earned on the investor's other options, then it would make no economic sense to undertake the investment. Only where the payoff to the investment, measured by the present value of expected future cash flows, exceeds the investment's cost will the project be unambiguously value-creating.

One significant qualification implicit in this discussion is that comparisons are being made among investments that share identical risk characteristics. Once we allow for the possibility that different projects have different risk profiles, determination of the project's opportunity cost is more complex. Since comparisons now have to be made on the basis of risk-specific rates of return, it is essential to devise some principle or set of principles capable of producing the correct hurdle rate for projects in different risk classes.

Modern finance theory has done just that by integrating two closely related ideas: (1) the mean-variance principle, namely, that risk and return are positively correlated – investors require higher rates of return on less certain investments to compensate for the higher risks they have to bear; and (2) the principle of portfolio diversification which says that so long as returns on different assets are less than perfectly correlated, investors will always be able to reduce volatility by building portfolios consisting of risky assets. In other words, the variance of the portfolio return will be lower than the volatility of the return on its constituent assets. The central insight of portfolio diversification is that risks that can be eliminated relatively easily and costlessly do not figure in the market pricing of risk. Only that component of risk which is unavoidable (what financial economists call, among other descriptors, systematic risk) is priced in the market.

These two principles are the foundation of the Capital Asset Pricing Model (CAPM), which asserts that required returns on risky investments contain three elements: a risk free rate, usually equated with the interest rate on government securities; the market premium, the difference between historic equity returns and the risk-free rate; and a parameter known as beta. With beta we come full circle since it is defined as the covariance between the expected return on a given asset and the market return divided into the variance of the market return. The greater the degree of independence between the return on a given asset and the market return, the lower will be the numerator and so too will the value of beta; beta in other words abstracts from diversifiable risks. All that one needs to know about the pricing of risk is contained in beta.

Things, however, are not quite that simple. Several empirical studies have investigated whether factors other than beta influence required returns. Indeed, multi-factor models have been developed that argue that risk can never be explained adequately in terms of a single parameter. These models predict required returns that differ, sometimes quite substantially, from those derived from the CAPM. But it would be rash given the present state of knowledge to assert the superiority of the former over the latter.

Project finance is by definition a highly leveraged transaction: borrowers attempt to finance the bulk of the project investment with debt. A question naturally arises whether investment values are or can be affected by the project's capital structure. A landmark study published in the late 1950s contained the 'proof' that under certain conditions the market value of the project or firm – a firm is a collection of projects – was independent of the way it was financed. Capital structure is irrelevant to the extent it does not influence the probability distribution of the investment's cash flow.

However, once it is recognised that most governments in effect subsidise debt by allowing firms to deduct interest expense for purposes of calculating income tax, it should be clear that financing choices, above all the use of debt, could have a direct and favourable impact on firm value. On this view, firm value is maximised by including only debt in the capital structure. While this prescription is theoretically impeccable, it fails to take proper account of bankruptcy or other risks that increase as the ratio of debt to total capital employed rises. Once such effects are factored into the analysis, a firm's optimal capital structure is determined as the point where higher bankruptcy and agency costs just offset the marginal tax benefits connected with the use of additional debt.

Chapter 3 extends the basic capital budgeting model by considering the ways firms actually undertake investments. The investment selection criteria discussed above appear far too static in terms of actual application. For example, it is nonsensical to imagine that investors quantify the payoff to a prospective investment solely in terms of base case economics. In the more usual case, investors will subject the investment to other forms of financial analysis, particularly sensitivity analysis, a technique that involves testing the robustness of a project's cash flow to alternative assumptions concerning the behaviour of different project variables. Sensitivity analysis provides no investment decision rules, only an evaluation of alternative states of key project variables capable of converting a favourable decision to proceed with an investment into an unfavourable one.

While sensitivity analysis goes some way toward redressing the static character of traditional investment analysis it cannot eliminate what others see as the fatal weakness of traditional methods. With or without sensitivity analysis, project economics are still evaluated as if nothing were to be gained from active management of project assets. The only circumstances under which it is legitimate to evaluate project economics using traditional methods is where the investment decision cannot be delayed or, alternatively, where it can be delayed but is reversible, meaning that investors are able to recover the project's original investment cost. Critics of traditional methods point out that the vast majority of investments fail to meet these criteria. In the more usual case, investments can be postponed, but once capital is committed they almost always are irreversible.

Taking the argument one step further, critics allege that traditional analytical methods fail to recognise the broad range of flexibility inherent in most project investments. Operators, for example, have the option to suspend production if market prices fail to cover cash operating costs, or the right to abandon projects if the present value of the expected payoff over the remaining life of the project is less than the project's current salvage value. Other equally compelling examples could be given that all point in the same direction: conventional capital budgeting methods do not in any way recognise the value of managerial flexibility. Such methods accordingly will always *understate* a project's true economic potential, meaning that too many profitable investments will be rejected. Since projects have option-like characteristics – or more to the point such characteristics can be created as part of project design – project evaluation should be based upon the same techniques, *mutatis mutandis*, used to value financial options.

Several studies have attempted to compare real options, as these characteristics have come to be known, with traditional methods to determine which approach more accurately replicates actual investment valuations. Unfortunately, the available evidence does not lead to the unambiguous conclusion that the options-theoretic approach produces valuations superior to those resulting from the application of traditional methodologies. Nor is there any compelling reason why they should. In many, perhaps most, instances the two approaches are complementary and can – and should – be used jointly.

Finally, there is the question of whether offshore investments should be evaluated applying the same principles and techniques used to assess the economics of national projects. Several analysts point to the fact that international investments may give rise to diversification benefits that will have the positive impact of lowering business risk. This means that required equity returns will be lower than those applying to undiversified firms, while internationally diversified firms should be able to source more lower-cost debt. The two effects together imply a lower-weighted average cost of capital and, correspondingly, a lower overall investment hurdle rate. On the other hand, offshore investments are subject to political risks, any action that might be taken by host governments, whether in the industrial or emerging market economies, that impairs project economics. Political risk ought to correspond to diversifiable risk: there is, for example, no real reason to suppose that a hostile act (say, expropriation) taken by one government will necessarily be mimicked elsewhere.

Financial markets are generally unwilling to accept such theoretical distinctions. The failure to ascribe any significance to international diversification stems from the premise that investors are unlikely to reward firms for doing for investors what they could just as easily accomplish for themselves and at lower overall cost. The only time investors will impute any value to diversification is when they are either unable to achieve similar results or because firms are capable of producing the benefit at lower cost. Research in the 1970s, for example, showed that share betas were inversely related to the proportion of a firm's total sales outside the United States; financial markets, in other words, rewarded US multinationals for having achieved lower business risks via diversification. However, once we recall that over much of this period the US government imposed controls on overseas portfolio investment to improve the nation's balance of payments, these findings are hardly surprising. Given the current high degree of international financial integration, it is doubtful whether this conclusion still applies.

Political risk is another matter. Numerous agencies produce country risk league tables. These indices are in the main concerned more with creditworthiness than with the sorts of risks that are of more immediate concern to foreign direct investors. Nor are bond premiums, spreads over US Treasury securities used to price international sovereign bond issues, any more useful since they, too, are concerned primarily with credit issues. If we accept that political risk is unavoidable in the sense defined above, then some mechanism must exist enabling investors to factor such effects into the derivation of appropriate financial benchmarks for international investments.

Several procedures have been proposed in the literature that do just that. Unfortunately, they start from fundamentally different conceptual premises, some favouring a total risk perspective while others rely on the CAPM framework. The principal distinction, of course, is whether international investors should take account of diversification benefits that may result from undertaking an offshore investment. The differences in terms of required equity returns are as might be expected enormous. Including diversification benefits results in project hurdle rates that in many instances are one half or less of those implied by the total risk approach. While the traditional approach is theoretically correct, the simple fact is most offshore investors use a total risk framework.

We have already indicated that project analysis and project financing go hand in hand: if an investment makes no economic sense to its sponsors, neither will it qualify for limited recourse financing. If, by contrast, sponsors are prepared to guarantee the debt that is another matter so long as they have 'deep pockets' to repay the loan. The same principles applied by project sponsors are used by commercial lenders to determine a project's economic viability. On the other hand, banks will typically use their own premises in respect of key project variables, output or input prices, for example; these premises may, but need not, correspond to those used by project sponsors. Given these premises, banks will develop a baseline estimate of project cash flow, which in turn is used to determine an investment's loan value. In some instances, lenders will constrain the size of the loan value by setting secondary credit tests designed to ensure that under virtually any state of the market outstanding loans will be repaid.

The more stringent requirements and higher costs associated with limited recourse finance raises the important question of why sponsors or host governments – which have come increasingly to rely on project finance to fund local infrastructure development – favour this form of borrowing over straight loans. Chapter 4 addresses this issue first by focusing on the interests of leading project participants, namely, project sponsors, host governments and commercial banks. Numerous theories have been advanced to explain the circumstances under which sponsors will choose limited recourse financing in preference to other debt options; we noted above some of the factors that might be expected to influence such decisions. Moreover, there is a large body of research that seeks to identify a single overarching motive for using limited recourse debt. A review of this literature leads to the conclusion that while it does provide useful insights into the importance of a number of factors that influence sponsor decision-making, it has yet to produce a fully convincing explanation of borrower motivation.

The issue appears even more paradoxical for infrastructure projects that historically were financed by the state out of the country's investment budget, general revenues or, alternatively, from the proceeds of offshore borrowings. The latter were typically arranged at spreads lower than those applicable to limited recourse loans and well below the returns required by private investors to undertake the same project. Moreover, host governments have an even broader menu of financing options than does the private sector. Such options range from privatisation and limited recourse funding, where project management and funding are left entirely in the hands of the private sector, to nationalisation where responsibility for such decisions lies entirely with the host government.

The obvious answer is also the correct one: the decision to use project finance to develop local infrastructure projects means that governments have come to accept the principle that resources should be valued at their true social opportunity cost, now almost universally equated with returns required by private investors. This, of course, is a relatively recent phenomenon. Previously, host governments, either among the industrial or emerging market economies, typically evaluated projects based upon an (arbitrarily set) official opportunity cost of capital, always assumed to be considerably lower than the return required by private investors (Dimson, 1989). The difference between the two was viewed as a measure either of private sector 'exploitation' or the extent of public subsidy. This is not to say governments no longer subsidise public sector projects. Of course, they still do. The key difference is that many no longer feel the need to conceal their intentions behind political or ideological notions.[13]

The correct answer as to why limited recourse finance is favoured over rival debt structures lies, in our opinion, in the extension of the insights produced by the real options literature. In general, the options-theoretic approach better explains most of the features that differentiate this form of borrowing from straight loans. The critical insight here is that debt, by increasing volatility, enhances prospective equity returns. It is hardly surprising, therefore, that investors opt for limited recourse debt since by definition it provides for a far higher degree of leverage than would be prudent were the project secured by the borrower's balance sheet. Project finance, in other words, is about risk shifting and not risk sharing. The latter, customary characterisation of limited recourse debt, derives from the complex web of contractual arrangements common to project finance. In fact, the options-theoretic view interprets such contractual arrangements in a fundamentally different way. Now they become the means by which banks ensure that loan proceeds are applied only to the very specific, narrowly-defined purposes described typically in minute detail in the loan agreement.

On either interpretation, risk remains very much at the centre of limited recourse financing. Chapter 5 defines project risks and looks at the ways they can be mitigated or eliminated; that is, it explores the means used by parties to the financing to address such risks and identifies third parties best able, in the sense of lowest cost, to provide the necessary coverage. The chapter also looks at the key features of project loan agreements, the risk management structure that protects lender interests. An understanding of the broad legal and credit principles underpinning this form of lending is essential to complete the analysis.

The chapter concludes with a brief discussion of project failure. Project failure is surprisingly rare, while abandonment of project assets in favour of the lending banks, permissible within a limited recourse framework, is rarer still. This is remarkable once we recognise that, statistically, projects are more likely to fail than succeed. True, this observation is based upon an analysis of natural resource projects, although there is no reason why this conclusion cannot be generalised to other industrial sectors.

Note, however, that such assessments are made *ex post*, that is, at the end of project life when investment performance is known with certainty. Investment decisions, however, are made *ex ante*, and expectations remain the driving force over project life. It should come as no surprise, therefore, that the options-theoretic approach again provides the most convincing explanation for the low rate of project failure. Most investors would prefer to stick with a project rather than forego potential upside – as they must given that the decision to abandon a project is irreversible – but only as long as such expectations are rational. We can easily imagine situations (the collapse of tin or crude oil prices in the mid-1980s, for example) that would lead to a fundamental reconsideration of investor expectations concerning the long-term price outlook and by extension prospective project returns. The fact that only very few projects were abandoned underscores the relevance of the options framework.

It also explains the reluctance of sponsors to hand over project assets to lending banks. Not too long ago credit considerations were seen to be the main source of resistance, the fear that, by defaulting, borrowers would be permanently shut out of the market. This possibility was always difficult to accept since borrowers not only choose limited recourse over other available options but also compensate lenders, in the form of higher spreads, for providing the option. It strains credulity that investors would select and pay the premiums associated with limited recourse finance without ever intending to exercise their legal option, except under the direst circumstances imaginable.

The final chapters present a series of case studies that illustrate the main issues presented in previous chapters.

Notes

1 Imagine that project cash flows are normally distributed. Krugman's point is that the project's cash flow potential should have been evaluated at the mean of the distribution rather than at the extreme right-hand tail, corresponding to the project's Pangloss value.

2 The calculation assumes that coverage ratios remained unchanged. If, however, government guarantees resulted in a reduction in coverage ratios, then maximum loan values would rise still further, to C^*/φ^*, where $C^* > C$ and $\varphi^* < \varphi$.

3 According to estimates compiled by the Institute of International Finance bank lending to five of the leading east Asian countries averaged $43 billion per year between 1994 and 1996. IIF forecasts are for net *outflows* of $27 billion in 1998 and a further $20 billion in 1999. According to the analysis from which these data are excerpted, 'idiotic foreign lending helped turn a property mania into an economic calamity' (Wolf, 1998) thus echoing Krugman's point.

4   Just over a decade ago commercial banks faced the same dilemma following the collapse of oil prices. The immediate (over)reaction was to withdraw from the market, as most major banks sharply reduced or eliminated their project finance departments (Smith and Walter, 1997). By the late 1980s the crisis had passed and banks this time were willing to finance infrastructure projects, which exposed lenders to risks different to, and potentially far more significant than, those that caused most lenders to reconsider their interest in this type of lending.

5   Lender interests are further protected by acquiring a charge over project assets. The value of such security depends upon the location of the project. For fixed project assets in emerging markets, the probability of being able to foreclose is obviously severely restricted. If the assets are movable, aircraft, for example, then once they depart from the defaulting country they can in most instances be seized. These observations highlight the defensive nature of security interests in project assets, that is, they establish the hierarchy of creditor claims in the event of default.

6   This distinction is what Wood and Vintner (1992) describe as *de facto* limited recourse, which they contrast with 'true' project finance debt.

7   'Project financing packages today are much more likely to have a capital market component than was true earlier. While straight bank loans remain the principal funding tool, an increasing number of projects are able to obtain their equity components in the Rule 144A market in the US or via domestic equity issues, and many larger and more creditworthy projects are able to tap international bond markets for supplemental capital' (Kleimeier and Megginson, 1997).

8   The proceeds of the $1.2 billion bond issue were used to construct the project's first train. Construction of a second train is covered by commercial loan financing, backed by export credit agency guarantees, in the amount of $1.6 billion. With the addition of the second train, senior project debt would rise from 46 per cent to 67 per cent of total capital, a figure that corresponds more closely to the sponsor's desired leverage (Rigby *et al.*, 1997).

9   The output of Ras Laffan is dedicated mainly to Korean Gas Corporation (Kogas). Potential gas demand was assumed to expand rapidly over the medium term. According to Kogas, Korean LNG demand prior to the recent financial crisis was forecast to rise from 9.2 million tonnes/year in 1996 to 20.7 million tones/year in 2001 and 29.3 million tonnes by 2010. The crisis has already had a negative impact on gas supply requirements. Kogas is reported to be considering cancelling 1998 spot and short-term contract LNG purchases compared with planned pre-crisis imports of 13.7 million tonnes. Part of this decline is connected with Kogas's decision to purchase a second Ras Laffan train. In exchange for higher purchases, Ras Laffan agreed to waive the floor price contained in the original sales and purchase agreement Thus contract supplies will displace a portion of spot and short-term imports, but will be priced competitively. Imports from Qatar under long-term contract are projected to total 0.6 million tonnes in 1998 rising to a steady state 4.8 million tonnes by 2005 (*Petroleum Review*, 1998).

10 Some support for this view is that while the project loan market contracted in 1997, project bonds, by contrast, rose by $1.6 billion or 54%. Most of the increase was accounted for by Latin America and, significantly, Asia. Total international bonds issued in the US market rose by about 7.5 per cent in 1997. By the fourth quarter of the year, Asian issues had all but dried up.

11 Rule 144A was passed in 1990 and such issues among other things are exempt from SEC registration and trading restrictions associated with private placements so long as trades are with Qualified Institutional Buyers (QIB). Nearly all

institutional investors qualify to purchase and trade debt issued under Rule
144A, creating a quasi-public market for this type of debt. 'To issuers, and
particularly international issuers, Rule 144A revolutionized the US market by
allowing access to a wide investor base while avoiding many of the disadvan-
tages associated with a public issuance' (HSBC, 1997).

12 The importance of non-traditional sectors is heightened by the fact that the
telecommunications industry now dominates US transactions (Sayer, 1997).

13 Not all governments are currently prepared to equate social discount rates
with those used by the private sector. In the important case of China, the lack of
any real progress in advancing private power projects is connected with the
government's refusal to allow foreign investors to earn returns commensurate
with project and political risks. To date, only a single foreign power project,
Laiban B – a joint venture comprising Electricité de France and GEC-Alsthom –
has been constructed; there is, significantly, some question whether the project
provides for an even remotely acceptable rate of return (Project Finance, 1998).
The obduracy of the Chinese government seems to derive from the traditional
practice of developing projects on a 'cost-plus' basis, which project sponsors,
with some exceptions, find unacceptable. Their view is that capital needs to be
fully costed. This naturally puts at risk planned generating capacity additions.
Official targets are for capacity to rise by 290 000 MW by 2000 and by an
additional 550 000 MW by 2010.

2 An Introduction to Capital Budgeting

Capital budgeting is concerned with how firms should evaluate projects and make capital investment decisions. Numerous factors enter into the formulation and implementation of corporate investment policy. These include such things as the cost and expected profitability of the project investment, determination of the investment's appropriate return, project risk management and mitigation, the project's optimal capital structure, the timing of the investment and the conditions under which it is economically rational to suspend or abandon rather than continue with project operations.

The numerous influences that shape project evaluation and selection can, in fact, be reduced to four basic propositions. The first of these is that financial decisions taken today have effects which are not realised immediately but rather are spread out over time. Second, since financial resources have alternative uses, robust decision criteria are needed to enable investors to order projects in terms of pay-off. Investors, in other words, need to be able to rank projects in terms of their value-enhancing potential. An obvious criterion would be to ensure that all eligible investments exceed prospective returns on the next best investment opportunity available to the investor; in economic theory this concept is known as the opportunity cost of capital. Opportunity cost, the third basic proposition, thus addresses two critical facets of financial decisionmaking: the scarcity of capital resources and, accordingly, the necessity for investors of having to choose among competing project options.

The fourth and final proposition is that cash flows received today are valued more highly than those that arise in the future. This principle is known as the time value of money. To capture the time value of money, future cash flows are discounted to their present value at a rate equal to the investor's opportunity cost of capital. And finally, once time becomes a critical input in decison-making, it is essential that investors are able to quantify exactly when costs and benefits arise, thus shifting the analytical focus from more traditional accounting concepts to cash flow. In cash flow analysis, costs are incurred when cash is actually committed and benefits are realised only when cash is actually received.

COMPOUNDING AND DISCOUNTING

Since discounting lies at the heart of capital budgeting, it is essential that the concept be defined precisely. At the simplest level of understanding, discounting is just the reverse of compounding. If we make an investment today at a known rate of interest, it is easy to calculate the prospective return one year hence according to the following expression:

$$FV_1 = PV_0 + rPV_0$$
$$= PV_0(1 + r),$$

where FV_1 is the value of the income to be received in a year's time, PV_0 is the initial value of the investment and r is the interest rate to be earned on the investment. If the time horizon of the investment is extended to, say, two years, then

$$FV_2 = FV_1 + rFV_1$$
$$= PV_0(1 + r) + r[PV_0(1 + r)]$$
$$= PV_0(1 + r)(1 + r)$$
$$= PV_0(1 + r)^2.$$

This process can be repeated over any number of time periods. Thus, the resulting general formula for the future sum of an initial investment equal to PV_0 earning r per cent per annum compounded over n years is:

$$FV_n = PV_0(1 + r)^n.$$

In n years time, the initial investment will be worth FV_n.[1]

Now, we can turn the question around and ask instead what in today's terms is an investment worth with a pay-off equal to FV in n year's time. To answer that question, all we need to do is solve the above formulas for PV instead of FV. In the one period example, the formula for the present value of an investment worth FV next year is:

$$PV_0 = FV_1/(1 + r).$$

The general formula can be expressed as:

$$PV_0 = FV_n/(1 + r)^n.$$

A sum payable in n years compounded at an annual rate of r per cent per annum is worth PV_0 today. In capital budgeting we are normally interested in evaluating the present value of a sequence of future payments rather than just a single payment. That is, we need to be able to evaluate the present value of a series of cash flows generated by an investment over its economic life. The expression for the present value of a series of future cash flows generated by an investment is:

$$PV = C_0 + C_1/(1 + r) + C_2/(1 + r)^2 + \ldots + C_n/(1 + r)^n,$$

or, more compactly, as

$$PV = \sum C_i/(1 + r)^i.$$

The present value of future income receipts is equal to the sum of the annual cash flows over n years discounted at an interest rate r. Expressed otherwise, discounted cash flow is simply a weighted average of future income, with the weights proportional to the relevant discount factor, $1/(1 + r)^i$. Note, too, that the summation is carried out from year zero to year n. Viewed this way, the initial investment, C, is in included in the calculation since, by definition, $C_0/(1 + r)^0 = C.^2$

The weights used to generate a cash flow profile are summarised in Chart 2.1, where three different discount rates, corresponding to interest rates of 5 per cent, 10 per cent and 15 per cent, are shown. The most obvious point is the higher the discount rate, r, the lower will be the present value of future cash flows. Indeed, the spread of the three curves provides a direct measure of the difference in present value resulting from the application of different discount rates. Note in this connection that a dollar or pound to be received in ten year's time will be worth today roughly four times more discounted at 5 per cent than if discounted at a 15 per cent rate.

Chart 2.1   Discount rate factors

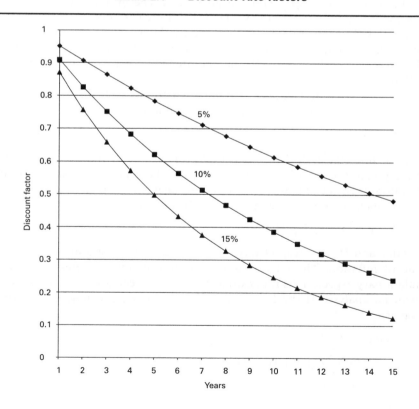

We can now pull together all the elements we have discussed to this point to arrive at a robust decision criterion for ranking and selecting individual investment projects, the so-called net present value (NPV) rule. The rule asserts that an investor should select all projects having an NPV > 0. NPV is defined as the present value of all future cash flows generated by the investment, net of the initial investment outlay. Since all future cash flows have been discounted at the appropriate opportunity cost of capital, a positive NPV signifies that the investment has generated value above and beyond what could have been achieved by the investor had the capital resources been committed to their next best use. The higher a project's NPV, the greater is its value-creating potential; however, any project commanding a positive NPV is worth undertaking. If a project's expected pay-off just equals the investment cost, that is, NPV = 0, the investor would be no better (or worse) off investing in the project than deploying the same resources in their next best use.

One interesting question that arises in this connection is whether cash flow potential should be evaluated in nominal or real terms; in other words, would the failure to take explicit account of inflation materially affect decisions reached based upon the NPV rule? The intuitive answer – that since inflation affects both the value of the future cash flows and the rate of interest used to convert them to their present value – is also the correct one, providing we assume a constant rate of inflation over project life and that financial markets have correctly incorporated the expected (constant) rate of inflation into nominal interest rates.

The first condition implies that nominal future cash flows (C) can be related to their 'real' counterpart (*C*) according to:

$$C_t = C_t(1 + \Pi^e)^t,$$

that is, nominal cash flows in period t are equal to the product of the corresponding real cash flow grossed up by the expected constant rate of inflation. If financial markets are efficient, then the nominal rate of interest (r) will include an inflation premium (Π^e), such that

$$1 + r = (1 + r)(1 + \Pi^e),$$

where r and Π^e are, respectively, the real rate of interest and the expected rate of inflation. The relationship connecting nominal interest rates and inflationary expectations is known as the Fisher effect, and is used widely in financial analysis. If the Fisher effect is correct, we can rewrite the NPV rule as:

$$\begin{aligned} \text{NPV} &= \sum C_t/(1 + r)^t \\ &= \sum C_t((1 + \Pi^e)^t/(1 + r)(1 + \Pi^e)^t \\ &= C_t/(1 + r)^t. \end{aligned}$$

If future cash flows are generated under the assumption of a constant expected rate of inflation, the NPV rule will yield the same answer whether the analysis is based upon nominal cash flows and nominal interest rates or real cash flows and real interest rates; the key point is the consistent application of the correct financial counterpart. Derivation of the real rate of interest, as would be required if the discounted cash flow analysis were developed using inflation-adjusted magnitudes, is accomplished by solving the Fisher equation for r. As a rough approximation, r is equal to the difference between the observed nominal rate of interest and the expected rate of inflation.

NPV is not the only project selection criterion and, if the available empirical data are to be believed, not even the one must widely used. Table 2.1 gives a number of alternative investment criteria, together with an assessment of their conceptual shortcomings, at least in comparison with the NPV rule. Two of the four alternative measures do not rely upon discounted cash flow methodologies, and thus fail to take explicit account of the opportunity cost of the capital tied up in the investment. In the case of the payback rule, it is possible to remedy the problem but only in part by replacing the undiscounted with the discounted cash flow series. This solution, however, does nothing to correct what is perceived to be the criterion's main shortcoming, namely that selection of the cut-off period is still arbitrarily determined.

In fact, most finance texts appear to be far too hard on the payback rule, since it can in fact be conceptualised as a variant of the internal rate of return (IRR), which is generally deemed a sounder and thus far more acceptable selection criterion. More to the point, a project's IRR is the discount rate that equates to a zero NPV; the payback rule, analogously, can be viewed as that rate which produces a zero NPV within the predetermined payback period. True, IRR values all project cash flows, while payback restricts the relevant cash flow stream to that arising solely within the cut-off period. On the other hand, neither is based on the correct investment rate – only NPV takes that rate explicitly into account – but rather compares the derived with the appropriate discount rate, the one that corresponds directly to the investor's opportunity cost of capital.

The use of discounted cash flow analysis is now widely recognised as the sole appropriate methodology for evaluating project economics. Numerous surveys have been undertaken over the past thirty years or more to quantify how widespread their use is in capital budgeting. Table 2.2, for example, summarises the results of a recent CBI survey (1994) that lists the financial analytical methods favoured by British manufacturing companies. The first point worth noting is that the overwhelming majority of respondents used, as expected, some form of quantitative analysis to assess project economics. However, of those firms that did, only about half used a discounted cash flow rate, the vast majority favouring use of the simple payback criterion.

Table 2.1 Alternative investment selection criteria

Measure	Definition	Comments
Payback	The number of years it takes before cumulative forecast cash flow equals the initial investment.	Selection of the cut-off period is arbitrary. The payback rule gives equal weight to all cash flows *before* payback and no weight at all to subsequent cash flows. The payback rule creates a tendency for the firm to accept too many short-lived and reject too many long-lived projects.
Discounted payback	The number of years it takes for cumulative forecast *discounted* cash flows to repay the initial investment.	Eliminates the equal weight cash flow problem before the cut-off date, but continues to ignore any cash flows that arise after that date.
Average return on book value	The ratio of the average forecast profit of a project (after depreciation and taxes) by the average book value of the investment.	By focusing on *average* returns no distinction is made for the fact that immediate cash flows are valued more highly than future cash flows. Thus, in contrast to the payback rule, which attaches no weight to future cash flows, book returns *give too much weight to such cash flows*.
Internal rate of return (IRR)	The rate at which the NPV of a project's cash flows are zero.	When an initial outflow is followed by a succession of inflows, IRR and NPV *give totally consistent signals*. Where projects are not independent or where the pattern of post-investment cash flows is not uniformly positive, IRR may fail.

Note, too, that a not insignificant minority applied an accounting rate of return notwithstanding the shortcomings inherent in the use of such a measure.[3]

Roughly two-thirds of the survey respondents evaluated project economics in nominal terms, although paradoxically the choice does not seem to make much difference to the value of the discount rate actually used; regardless of which approach was favoured respondents applied a discount rate of around 16.5 per cent. This, of course, means that the implied nominal discount rate used by firms that project cash flows in constant terms is higher than the rate applied by respondents favouring the use of current magnitudes, by as much as 500 basis points if the expected inflation rate given in the survey is to be believed. Why firms that evaluate project cash flows in real terms should apply a higher discount rate than those that do not is by no means obvious.

Table 2.2   How do British firms evaluate projects?

Question	Per cent
Undertake quantitative assessment	90
Of those answering affirmatively, the methodology employed was:	
• an accounting rate	13
• a discounted cash flow rate	53
• simple payback	75
• return on capital	49
• return on equity	12
• other	3
The required rate of return is calculated on a:	
• pre-tax basis	56
• post-tax basis	41
The required return is based on:	
• real figures	35
• nominal figures	63
For those companies that use real figures, the mean is:	16.4
For those companies that use nominal figures, the mean is:	16.8
Those companies using nominal data expect the average annual inflation rate to be:	4.9
Sensitivity analysis is employed	48

Source: CBI (1994).

The importance of payback also emerges in a recent survey of the capital budgeting practices of American multinationals. Respondents to the survey indicated that payback is the preferred method for adjusting the discount rate to take account of project risk, closely followed by subjective criteria. Indeed, there is clear evidence that American multinationals have long favoured such criteria (Buckley, 1996a), suggesting either that the subjective element is far stronger in offshore than in domestic investments or, less plausibly, that British firms are more financially sophisticated than their American counterparts.

Finally, we might ask whether investment decisions actually follow the NPV rule. This question is less absurd than it might at first appear since there is considerable evidence of a close association between investment expenditure and current cash flow. This relationship clearly contradicts the standard NPV rule, which asserts that expected profitability, as measured by the present value of future cash flows, should be the critical determinant. Unless the former is a lead indicator for the latter, a not entirely unreasonable supposition, investment spending should be independent of the level of current profitability.

The correlation between capital spending and current cash flow is illustrated in Verleger (1994). This chart compares the level of investment expenditure by the world oil industry with the price of crude oil, used here as a surrogate for industry profitability. The results for this one sector can be generalised to the economy as a whole. Blanchard (1997), for example, provides graphic evidence of how close the relationship between the two series has been for the American economy over the 1960–94 period; he also discusses the results of recent empirical studies that appear to show the superiority of current over expected profitability as a leading determinant of investment spending.[4]

One possible explanation for these findings is that the NPV rule is not wrong, only incomplete. More to the point, it is possible to argue that investment decisions are determined in the light of available financial capacity. Since internally generated funds are a major source of investment financing, a decline in cash flow must result in a decline in investment outlays. Following this line of reasoning, we might expect investment spending to respond to both current and expected profitability.

Thus we have three competing explanations for an observed empirical fact. The first is that investment decisions are driven by positive NPVs. Unfortunately, the data needed to construct an economy-wide measure of NPV simply do not exist. However, the so-called Tobin q-ratio – the ratio of the market value of firm assets to their estimated replacement cost – provides a close approximation. In this formulation, it is worthwhile undertaking an investment if $q > 1$; if $q < 1$, the expected pay-off from the investment is lower than its cost, the investment should be abandoned and the existing capital stock reduced until $q = 1$. Put another way, at the margin the cost of an investment must equal its expected benefit, which in turn depends upon the difference between q and 1. The numerator of the q-ratio includes the market value of both debt and equity, while the denominator includes all assets, not just net worth. Measurement at replacement cost has the additional advantage of eliminating any biases that might result from past rapid rates of inflation. An alternative formulation of the importance of expectations starts from the premise that profit prospects should be reflected in share price movements. In which case, investment spending should bear a close relationship to changes in stock market performance.

The second hypothesis asserts that cash flow variables are the most significant determinant of investment spending since it is impossible to separate financing from investment decisions. The third hypothesis allows the level of investment expenditure to be co-determined by current and expected profitability. Note that only the first hypothesis and its stock market variant are consistent with the proposition that investment decisions are driven by profit expectations. The other hypotheses, by contrast, pose a serious challenge to accepted capital budgeting theory.

Barro (1990) presents quantitative evidence that permits evaluation of the importance of each of the three hypotheses. These results, presented in the following table, provide persuasive evidence of the superiority of expectational over other effects: in every case, regressions containing the stock market variable outperform rival models. The first equation shown in the table corresponds to the pure NPV hypothesis. True, q is statistically significant but note how poorly the equation performs, especially in comparison with equation 1a, which has a higher R^2 – signifying that it 'explains' more of the variance in fixed non-residential expenditure – and a lower standard error of estimate.[5]

Table 2.3   Equations for fixed US non-residential private investment, 1947–87

Hypothesis 1: Investment is driven by positive NPVs

$$DI = 0.031 + 0.065DI_{-1} + 0.28Dq_{-1}$$
$$R^2 = 0.29 \quad SE = 0.59$$

Hypothesis 1a: Investment is driven by profit expectations proxied by share prices

$$DI = 0.01 + 0.347DI_{-1} + 0.298S_{-1}$$
$$R^2 = 0.42 \quad SE = 0.52$$

Hypothesis 2: Investment is driven by cash flow availability

$$DI = 0.33 - 0.026DI_{-1} + 3.42D\Pi_{-1}$$
$$R^2 = 0.13 \quad SE = 0.65$$

Investment is driven by both expectations and Tobin's q

$$DI = 0.014 + 0.302DI_{-1} + 0.25S_{-1} + 0.077Dq_{-1}$$
$$R^2 = 0.43 \quad SE = 0.53$$

Hypothesis 3: Investment is driven by both expectations and cash flow

$$DI = 0.013 + 0.271DI_{-1} + 0.276S_{-1} + 2.18D\Pi_{-1}$$
$$R^2 = 0.47 \quad SE = 0.52$$

Variables used in the analysis

Variable	Description
DI	Growth rate of investment.
S	Growth rate for year t of the real stock market index. NYSE returns in nominal prices less the GDP deflator.
DΠ	The first difference in the ratio of after-tax corporate profits to GDP.
Dq	Growth rate of q, where q is the ratio of the total nominal value of non-financial corporations to the capital stock at nominal replacement cost.

Source: Barro (1990).

Note: Non-italicised variables denote statistical significance at the 95% or higher level.

Equation 1a captures expectational effects via changes in share prices. Positive changes in equity prices imply an improvement in profit potential which, with a one year lag, affect investment spending. In other words, the implied improvement in profitability increases the expected pay-off from the investment by increasing its value in relation to the original investment cost thus raising the investment's NPV. Note, too, that the stock market variable dominates q. Where the two variables appear in the same equation (4), the latter loses statistical significance and the resulting fit is little different from the equation containing just share prices. Finally, the cash flow model is the poorest of the lot, explaining less than 15 per cent of the variance in investment spending over the sample period. Nor is the addition of the stock market variable able to salvage the importance of cash flow as a determinant of investment spending: current profits are only marginally significant and again the result shows no improvement over equation 1a.

Despite the conviction that current cash flow is a better lead indicator of investment expenditure than is expected profitability, the evidence suggests just the opposite conclusion. One interesting question is why the stock market variable should be superior to q. After all, q includes changes in equity values and thus ought to perform at least as well. Barro's explanation is that the growth rate of q tends to be dominated by changes in equity values; the correlation between the two series is 0.94. Furthermore, changes in equity values are considerably more volatile than are fluctuations in either net debt or the replacement cost of capital. In other words, whatever significance there is in changes in q can be attributed entirely to fluctuations in stock market values, hence the superiority of equation 1a over equation 1.

An alternative explanation is provided by Dixit and Pindyck (1994). Analysing investment decisions within an options as opposed to a neo-classical framework, they argue that the selection criterion implied by the q-ratio, invest when $q > 1$, is not sufficiently rigorous as it excludes the value inherent in being able to adjust the timing of the investment decision to improved market conditions. On this view, the relevant decision criterion is not $q > 1$, but rather $q^* > 1$, where $q^* > q$ and incorporates the benefit derived by investors from being able to postpone an irreversible investment. By extension, the same criticism applies in respect of standard NPV calculations.

PORTFOLIO BUILDING

Our discussion thus far has treated project outcomes as if they were known with certainty. In practice, the cash flow profile upon which investment decisions are based derives from the base case, invariably the most likely of a set of possible outcomes. Alternatively, a project's NPV may be viewed as a weighted average of all possible financial outcomes, with the weights proportional to the probability of any individual outcome occurring. Potential returns can thus be described by a probability distribution, with the most likely outcome corresponding to the mean of the distribution, while the spread or variance of possible outcomes can be determined with reference to the standard deviation of the same distribution. The larger is the standard deviation, the less certain is the outcome; the potential returns to the investment, in other words, are subject to greater risk.

One of the commonplaces of modern financial economics is the conviction that a positive relationship links risk and return: the riskier the pay-off to a given investment, the higher the required return will have to be to induce an investor to undertake the project. Risk, however, is a multi-layered concept so it is essential that each layer be properly understood, all the more so since financial markets price only that component that is unavoidable, that is, that portion of the total risk inherent in an investment that cannot be eliminated easily and relatively costlessly. This is the critical insight of portfolio diversification.

A portfolio is a collection of assets that is specifically constructed with the objective of achieving the highest possible return for a given level of risk. The recognition that returns from different assets do not necessarily move together is central to portfolio building; in most instances, the covariance is less than perfect. This property is crucial to understanding the contribution diversification can make toward risk reduction. More to the point, by combining two or more assets whose returns are largely independent, the combined portfolio return will be less volatile than are the returns on the constituent assets.

To see this, we begin first by noting that the expected return on a two asset portfolio is equal to the sum of the underlying returns weighted by their share in the portfolio. More formally, the portfolio return, r_p, is given by:

$$r_p = w_j r_j + w_k r_k,$$

where r_j is the return on asset j, r_k is the return on asset k and w_j and w_k are the shares of assets j and k, respectively, in the overall portfolio. The variance of the two asset portfolio is defined as:

$$\sigma_p^2 = w_j^2 \sigma_j^2 + w_k^2 \sigma_k^2 + 2 w_j w_k \sigma_{jk}$$

where σ_p^2 is the portfolio variance, σ_j^2 and σ_k^2 are, respectively, the variances of returns on assets j and k and σ_{jk} is the covariance between j and k. By definition, the correlation between j and k can be written as:

$$\rho_{jk} = \sigma_{jk}/\sigma_j \sigma_k,$$

so,

$$\sigma_{jk} = \rho_{jk} \sigma_j \sigma_k.$$

If this expression is substituted for σ_{jk} in the variance formula, we get:

$$\sigma^2 = w_j{}^2\sigma_j{}^2 + w_k{}^2\sigma_k{}^2 + 2w_jw_k\rho_{jk}\sigma_j\sigma_k.$$

Now, if the returns of assets j and k are perfectly correlated, that is, $\rho_{jk} = 1$, the variance of the portfolio return is equal to the weighted average variance of the underlying assets:

$$\sigma_p{}^2 = (w_j\sigma_j + w_k\sigma_k)^2$$

and

$$\sigma_p = w_j\sigma_j + w_k\sigma_k.$$

If, by contrast, $\rho_{jk} = -1$, the portfolio variance equals zero, and is achieved by setting the portfolio weights as $w_j = \sigma_k/\sigma_j(1 + w_j)^{-1}$ and $w_k = (1 - w_j)$. Where the returns are perfectly positively correlated there are no benefits connected with diversification; where the returns are perfectly negatively correlated, all of the risk can be eliminated from the portfolio. In the more usual case, the covariances among the returns of the underlying assets determine the magnitude of the benefits to be obtained from holding a diversified portfolio. It should be stressed that while the analysis has been developed as if the portfolio consisted solely of financial assets, the results are perfectly general and apply with equal validity to project portfolios.

The results derived from consideration of the two asset portfolio can be generalised although the computations become increasingly complex as the number of assets included in the portfolio increases. In general, as the size of the portfolio increases, the portfolio variance approaches the average covariance. If all assets in the portfolio were uncorrelated with each other, the portfolio variance would fall to zero; all risk, in other words, would have been eliminated from the portfolio However, since most assets tend to be positively correlated, the resulting positive covariances set a limit to the benefits of diversification (Brealey and Myers, 1991). As a general matter, risk declines sharply as the number of assets in the portfolio approaches ten; thereafter, the incremental benefits are much smaller (Statman, 1987).

Finally, Figure 2.1, taken from Bernstein (1996), illustrates how powerful is the contribution that diversification can make towards risk reduction. The plot shows the risk–return characteristics of investments in stock markets in 13 emerging countries, the same trade-off for investments in the US S&P 500 Index, and a portfolio consisting of each of the 13 emerging stock markets. Owing to the fact that individual returns are largely uncorrelated, the standard deviation of the combined emerging market portfolio is actually lower than the standard deviation of any of the returns on the constituent indices.

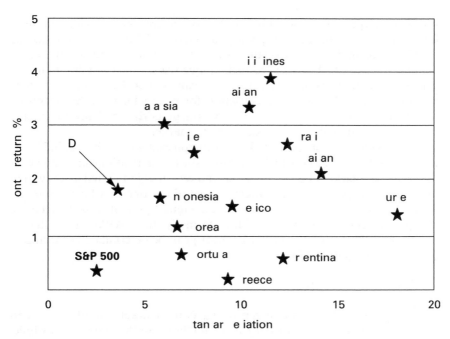

Figure 2.1 Portfolio diversification

Source: Bernstein (1996).

THE CAPITAL ASSET PRICING MODEL

The central conclusion of the preceding discussion is that risk can be reduced but cannot be eliminated entirely. To build a zero variance portfolio requires finding the right combination of assets whose returns are perfectly negatively correlated. It is extremely unlikely it will ever be possible to do so since all returns, to a greater or lesser extent, fluctuate with the level of overall economic activity. The residual risk remaining in a fully diversified portfolio derives precisely from these positive covariances. This risk is known as systematic risk, defined as that component of total risk that cannot be eliminated through diversification. Unique risk, by contrast, results from developments specific to the firm. It is unrelated to the state of the economy and thus can be eliminated by building portfolios whose returns are more or less unrelated.

Consider the following example which should help to clarify the point. We know that mineral exploration is a high risk activity: for instance, only around 1 in 10 exploratory wells drilled will be successful. However, the success or failure of an individual energy company to find commercial reserves is largely independent of the state of the economy; it relates more or less solely to technical and geological factors. Thus, even though the likelihood of finding oil is small, an investor holding shares in an energy company need not be exposed to such unique risks, which can be eliminated via diversification. Accordingly, investors can neither demand nor are they entitled to any additional compensation for risks which can be addressed within the context of their overall investment portfolio. Only systematic risk, which is unavoidable, is priced in the market.

The pricing of systematic risk is accomplished using the capital asset pricing model (CAPM), which establishes a linear relationship, known as the securities market line, between risk and return of individual assets within in an efficient portfolio. An efficient portfolio is one in which assets are combined in such a way so as to yield the highest possible return given the risk preferences of the investor. More formally, the CAPM shows that in equilibrium the return, r_j, on a risky asset j can be determined according to the following relationship:

$$r_j = r_f + \beta_j(r_M - r_f),$$

where r_f is the riskless rate of return, r_M is the market rate of return and β measures the volatility of asset j's return relative to the market as a whole, and

$$\beta = \sigma_j \rho_{jM} / \sigma_M.$$

In words, beta is defined as the ratio of the covariance between asset j and the market return to the standard deviation of the market portfolio. The smaller the covariance is between the return on a given asset relative to the market, the smaller will be the resulting value of beta, meaning the lower will be the return required by the investor for including asset j in his or her existing portfolio. Beta, in short, abstracts from unique risk, pricing the incremental risk associated with the addition of asset j to the investment portfolio solely in keeping with its systematic risk.[6]

For present purposes, the value of the CAPM lies in its ability to quantify the premium required by an investor for undertaking a risky investment. In our initial discussion of project selection, it was argued that investments can be ranked according to their respective NPVs and it is worthwhile under-taking all projects having a positive NPV since each will be value-enhancing. Risk-averse investors will always impute a higher required return to invest-ments whose pay-offs are less certain, such that riskier projects will have to command a premium over less risky investments if they are to be undertaken at all. One way to accomplish this would be to penalise the riskier investment by increasing the discount rate used to convert prospective cash flows to their present value equivalents. This is conceptually akin to adding a risk premium (r_P) to the discount rate used for evaluating risk free cash flows:

$$r_j = r_f + r_P.$$

Since from the CAPM $r_P = r_j - r_f = \beta(r_M - r_f)$, substituting we obtain:

$$r_j = r_f + \beta(r_M - r_f),$$

which says the premium required by an investor for undertaking a risky investment is equal to some multiple of the market premium, with the size of the multiple determined by the magnitude of beta.

The application of the CAPM to capital budgeting is fraught with difficulty. In practice project betas do not exist so that investors perforce have to rely on share betas as a surrogate. The closer the fit between the proxy company and the project under consideration, the better will be the match between the observed beta and the true opportunity cost of the project it purports to measure. However, firms obviously differ in terms of technical sophistication and competence, management style and the extent to which their operations are diversified across different industrial sectors. As might be expected, individual share betas tend to average out all of these influences.

Moreover, as Rosenberg and Rudd (1998) have argued, share betas used to calculate required project rates of return suffer from several additional shortcomings, the most significant being that the choice of comparator is arbitrary. This limitation, when 'combined with the fact that the number of similar companies is often very small, restricts the number of cases where (the "method of similars") can be used with confidence.' Rosenberg and Rudd go on to argue that a far more satisfactory methodology would be to use what in their terminology is called a fundamental beta derived, unlike share betas, directly from traditional accounting information, balance sheet or income statement data, for example. The superiority of fundamental over share betas lies in their alleged greater accuracy, more satisfactory method of derivation and direct connection to the operating characteristics of the company, division or project to which it relates.

Among the variables used by Rosenberg and Rudd to estimate fundamental betas is the degree of financial leverage, which has the same significance for more traditional company betas. The key point here is that the higher the debt to equity ratio is, the higher will be the resulting estimate of beta and hence the required rate of return. This follows from the fact that debt represents a fixed claim on the project's cash flow, so that any decline in earnings will result in a corresponding decline in the amount of cash available for distribution to shareholders. With leverage, more of the financial risk of the investment is reserved for the equity holders, who will thus require greater compensation in the form of a higher required return for having to bear it. Now, unless the capital structure of the project corresponds identically to that of the proxy company, the estimated required rate of return will be biased upwards inviting the risk of rejecting an investment that otherwise would have been accepted. It is, however, possible to unlever an observed (asset) beta, and in so doing to quantify the impact that alternative levels of debt will have on required project returns.

To disaggregate the observed asset beta into its constituent parts, we begin first by assessing the impact the use of debt has on required returns. The following formula is the general CAPM expression for calculating required returns in the presence of leverage:

$$r_j = r_f + \beta_U[r_M - r_f] + \beta_U(1 - \tau)D/E[r_M - r_f].$$

The relationship says the required return on a leveraged investment is equal to the sum of (1) the risk-free rate, (2) the pure equity return, defined as the product of the market premium times the unleveraged beta, and (3) an incremental return equal to the market premium multiplied by the firm's beta grossed up to take account of its current debt to equity ratio (D/E) adjusted by its current tax rate (τ). This latter offset arises from the fact that firms can deduct interest payments for tax purposes. The same expression can be written alternatively as follows, which provides a more explicit description of the impact of leverage on required returns:

$$r_j = r_F + \beta_U[r_M - r_f] + \beta_L - \beta_U[r_M - r_f].$$

With no debt in the capital structure, the expected equity return to an investment is simply the sum of the first two terms; the effects of leverage are captured entirely in the third component. Within a CAPM framework, the financial risk component is simply the increase in beta, $\beta_L - \beta_U$, caused by financial leverage, multiplied by the risk premium on the market as a whole, $r_M - r_F$. Additional debt amplifies the systematic risk inherent in a firm's business operations increasing both beta and the expected return.

By the first expression above, we know that the leveraged beta is equal to

$$\beta_L = \beta_U[1 + (1 - \tau)D/E],$$

while the unleveraged beta can be written as

$$\beta_U = \beta_L/[1 + (1 - \tau)D/E].$$

Given the firm's current debt/equity ratio, tax rate (τ), and the value of the observed leveraged beta we can unlever the beta by solving the second equation to derive β_U. Substituting this value together with the new values of D, E and τ into the first expression we can derive the appropriate value of beta consistent with the degree of leverage in the contemplated investment project. The estimated beta associated with the new debt ratio is then inserted into the securities market line to derive the required return associated with the new D/E ratio.

The popularity of the CAPM derives principally from its strong intuitive appeal and the fact that the securities market line provides a precise numerical estimate of the return required to undertake an investment. All the information an investor needs to assess the risk of different investment options is contained in beta. Recent research, however, has cast considerable doubt on whether things really are that simple: if these results are to believed beta provides a poor index of project risk. No other factors, either individually or in conjunction with beta, have yet been identified that would remedy the problem; this inability, too, goes a long way toward explaining the continuing support for the CAPM.

To put these negative findings in their proper context we need first to review the assumptions underpinning the CAPM. The first critical assumption is that investors share a common view of prospective returns, variances and covariances. This assumption implies that all the relevant information one needs to know about an investment is contained in the mean and standard deviation of the asset's return. The second key assumption is that the capital market is in equilibrium. Markets, in other words, are assumed to be efficient in the sense that current prices contain all presently available information concerning the investment. Seen this way, the CAPM can never be tested directly since any empirical analysis is ultimately a test of the theory's constituent parts. If a test of the CAPM fails, it is impossible to know for sure whether the failure lies with the theory itself or with the possibility that the market in which the asset is priced is inefficient. And finally, the CAPM is a one period model meaning that prospective returns are assessed over the next period only. If the investment horizon of different investors is not the same, we might expect their behaviour to differ as each seeks in their own way to maximise the benefit of their wealth. A simplified model that abstracts from such differences can never adequately describe the resulting risk–return tradeoff.

Enumeration of the principal assumptions underlying the CAPM highlights many of the practical problems that have been encountered in testing the theory's main conclusions. The CAPM is ultimately concerned with expected returns. But as investor expectations are unobservable, tests of the CAPM have typically utilised actual returns, a perfectly acceptable alternative providing investors hold to the same view of the process generating the realised returns. This is not too strong an assumption since we know that as long as some investors form their expectations rationally then the market as a whole will behave rationally (Pollio, 1985). If correct, then tests of the CAPM are well specified since the joint hypothesis, that markets are efficient, has similarly been addressed. A far more serious difficulty with using actual returns is that the underlying data are subject to considerable noise – financial markets are continuously affected by surprises that result in very high observed volatilities. A high noise to signal ratio could obscure the underlying relationship resulting in the erroneous rejection of the theory. Also, by definition the market portfolio contains all risky assets. Tests of the CAPM, however, typically rely on a subset of financial assets, usually stocks traded on the NYSE, the AMEX or NASDAQ, again possibly biasing the results against the theory.

All of these considerations should not be taken to mean that empirical tests of the CAPM are worthless. Quite the contrary: if the main conclusions of the CAPM are verified in spite of these problems, there can be no question of the theory's robustness. The real dilemma lies in the failure of empirical analysis to verify the main predictions of the CAPM since such findings can never be accepted uncritically. In which case, the ultimate verdict on the CAPM must depend upon both the quality and consistency of the evidence.

The dilemma is all the more significant as the bulk of recent empirical evidence is strongly against the main predictions of the CAPM. Most analysts have approached testing the CAPM by constructing a number of stock portfolios sorted according either to size as measured by market capitalisation or by beta as a measure of risk. If the CAPM is correct, there should be no systematic relationship between size and realised returns since that would imply that some factor other than beta affects the market pricing of risk. Similarly, the CAPM predicts that if portfolios are sorted by beta, those having higher risks should command a premium over lower risk portfolios, with the premiums corresponding identically to those described by the securities market line.

Table 2.4 provides evidence on each of these points by comparing the relationship between firm size, achieved by grouping individual companies into ten portfolios on the basis of market capitalisation, and realised returns for stocks listed on the New York Stock Exchange for the period 1926 to 1994; average betas are also shown for each decile. The results with respect to the CAPM are mixed. First of all, there is a strong relationship between size and actual returns: over the past 70 years, realised returns for firms in the tenth decile, the smallest in the sample, were 400 basis points higher than those for the largest NYSE quoted companies. On the other hand, there is, as expected, a positive relationship between actual returns and beta although the evidence indicates that, with the exception of the very largest firms, the CAPM systematically understates actual returns with the margin of error inversely related to firm size.

Several analysts have tried to rationalise the importance of size effects without apparently fully appreciating the difficulties such explanations, however convincing they may be, pose for the CAPM. If the CAPM is

Table 2.4 Firm size and realised returns for common stocks listed on the NYSE during the period 1926–94

	Decile	Beta	Mean return	CAPM-implied return	Size premium
Largest	1	0.90	11.01%	11.45%	−0.44%
	2	1.04	13.09	12.46	0.63
Mid-Cap	3	1.09	13.83	12.82	1.01
	4	1.13	14.44	13.11	1.33
	5	1.17	15.50	13.24	2.16
Low-Cap	6	1.19	15.45	13.50	1.96
	7	1.24	15.92	13.87	2.05
	8	1.29	16.84	14.17	2.67
Micro-Cap	9	1.36	17.83	14.69	3.14
	10	1.47	21.98	15.45	6.53
Mid-Cap		1.12	14.32	13.01	1.31
Low-Cap		1.23	15.87	13.75	2.12
Micro-Cap		1.39	18.92	14.90	4.02

Source: Emery and Finnerty (1997).

correct, these effects should not exist. The fact that they do implies that at best the CAPM provides an incomplete explanation for the pricing of risk or at worst is wrong. In this connection, Emery and Finnerty (1997) argue that size is a perfectly plausible proxy for risk. 'Larger firms tend to have more information available about them, to have longer operating histories, to be more well established and to be better diversified (across product lines, geographical regions, and so forth) than smaller firms.' On the other hand, they go on to note that without an asset pricing model that specifies the precise nature of the relationship between size and risk, its existence must remain little more than an empirical curiosity.

The evidence given in the table provides qualified affirmative support for the main implications of the CAPM. If the theory were indeed robust, we should expect the indicated positive relationship between realised returns and beta to hold up over any sub-periods that fall within the full sample period. This, however, does not appear to be the case. Fama and French (1992) and, more recently, Malkiel and Xu (1997) show that since the mid-1960s the relationship between actual returns and beta no longer holds. The securities market line has become perfectly flat: high risk portfolios as measured by beta now command premiums no greater than do lower risk portfolios. Furthermore, Fama and French find that the ratio of book value to the market value of a firm's equity helps to explain the cross-sectional variation in average stock returns while both Fama and French and Malkiel and Xu reaffirm the importance of size effects. The latter study, however, suggests that size may be a proxy for unique risk given the strong inverse relationship they find that exists between the two variables. These results, if correct, damn the CAPM twice over. First, they again show that factors other than beta affect realised stock returns and, second, that in terms of the pricing of risk the market does appear to take into account unique risk or whatever other factors their measure actually represents.[7]

These findings have not gone unchallenged. Both Chan and Lakonishok (1993) and Black (1993), among others, provide detailed critiques of the Fama and French study. The irony is that both of these articles, albeit in different ways, actually play into the hands of those whose results they ostensibly seek to rebut. The Black study, for example, provides strong evidence that the relationship linking actual returns with beta has changed significantly since the mid-1960s. If we plot the market premium against beta for the period 1966–91 and compare it with the same plot for the full sample period and the 1931–65 sub-period, something very wrong is apparent. Risk-seeking investors should expect to earn higher returns compared to risk-averse investors and that is precisely what the plots for the full sample period and the 1931–65 sub-period show. Only in the most recent sub-period does the market fail to compensate risky portfolios with higher premiums. Black interprets these results as reflecting nothing more than that low beta portfolios did about as well over this period as did high risk portfolios, surely a contradiction in terms of the CAPM.

Similarly, Chan and Lakonishok assert that the evidence supporting the case for beta could just as easily be turned around and used against it. Their verdict on the CAPM is that 'the data simply do not lend themselves to a clear-cut conclusion either way ... it may very well be the case that returns are driven not only by risk but also by a host of other institutional and behaviourial aspects of equity markets.' The unique Scottish verdict they seem to favour – not proven – looks more like an attempt to hold on to a theory that has passed its sell-by date than a reasoned justification for its continued use.[8]

The CAPM is only one, albeit the best known, of several asset pricing models that can be used to determine the required rate of return on a risky asset. The CAPM's main rival is the arbitrage pricing theory (APT), which starts from the same premise as the CAPM, namely, that investors are compensated only for bearing systematic risk, but then goes on to quantify systematic risk in a totally different way. Like the CAPM, APT assumes that financial assets vary in their sensitivity to macroeconomic risks but takes the view that beta is too crude an index to capture such effects. Instead, it posits a relationship in which risk is measured by a set of macroeconomic factors that collectively determine the required return on a financial asset.

More formally, the risk premium under APT is determined according to:

$$r_j - r_f = \phi_0 + \phi_1(F1 - r_f) + \phi_2(F2 - r_f) + \phi_3(F3 - r_f) + \ldots + \mu$$

where r_j is the return on asset j, r_f is the risk free rate, $(r_j - r_f)$ is the risk premium to be earned by asset j, F_i are the macroeconomic factors that affect r_j, ϕ_i are parameters that measure the sensitivity of the expected return to the respective factors. The last term, μ, is analogous to unique risk in the CAPM: it measures events that are specific to the firm and thus can be eliminated via diversification.

According to the APT, a portfolio that is structured so as have zero sensitivity to the individual macroeconomic factors will be priced at the risk-free rate. If the actual return exceeds r_f, the investor could earn a risk-free profit, which is what arbitrage means, by borrowing at the risk-free rate and buying the portfolio; if, by contrast, the portfolio earned less than the risk-free rate, just the reverse strategy would be pursued. Arbitrage thus ensures that the portfolio earns exactly the appropriate rate of return for its risk class. The second feature of the model is that the required return on any asset is directly proportional to its sensitivity to each of the macroeconomic factors. If, for example, asset j were twice as sensitive to F1 than asset k, with all things constant, the expected risk premium on j would be twice as large as that applying to k.

Note, too, that the market portfolio, which is central to the CAPM, plays no role at all in APT. Furthermore, instead of relying on a portmanteau variable (beta) to capture all non-diversifiable risk, the APT seeks to define a specific set of macroeconomic variables that relate directly to required returns. The Malkiel and Xu (1997) results discussed above underline the potential significance of this feature of the APT. Their measure of unique risk could just as easily be rationalised as capturing the sorts of influences that the APT seeks to quantify. On the other hand, the main shortcoming of the APT is that it fails to define the set of macroeconomic factors to which required returns relate. True, Chen *et al.* (1986) have attempted to isolate the relevant factor set. Their analysis suggests the importance of the level of industrial activity, the rate of inflation, the spread between short- and long-term interest rates and the spread between the yields on low and high grade corporate bonds. But since the factor set does not emerge directly from the theory itself, the APT is susceptible to the charge of data mining, precisely the sort of criticism widely levelled against those studies that have attempted to discredit the CAPM.[9]

The CAPM and APT are both fraught with difficulties that impair their usefulness for determining appropriate equity returns on risky assets. If so, the main implication must be that the determination of risk premiums is more art than science. Certainly, surveys of capital budgeting practices used by leading industrial or multinational firms, whether British or American, seem to bear this out. We leave until later (Chapter 3) consideration of how firms approach the problem of asset valuation. For now, however, we might note that several studies advocate use of what has come to be known as the total firm risk approach for implementing capital budgeting decisions (van Horne and Wachowicz, 1995). This methodology takes as its starting point the fact that individual firms can build project portfolios that not only add shareholder value but have the knock-on effect of reducing firm risk. Firms, in other words, search out potential projects whose cash flows are largely uncorrelated with existing assets in order to exploit the benefits of diversification. The contribution of one or more potential projects is evaluated in terms of the relative impact their implementation would have on total firm risk. The decision criterion is thus to accept all projects having positive NPVs *and* which reduce firm risk.

Projects that fail to meet the second part of the criterion, even though by definition they are patently value enhancing, are rejected. Therein lies the main shortcoming of this approach: projects that are potentially capable of creating shareholder wealth are passed over in favour of those capable of producing diversification benefits, benefits that could just as easily be accomplished at lower overall cost by the firm's shareholders. Whatever intuitive merits the total firm risk approach may have, it should be noted that the wave of conglomerate mergers that occurred in the 1960s was motivated by precisely this objective. These mergers were subsequently unwound in the 1980s as investors came to recognise that the whole was in fact worth less than the sum of its parts.

CAPITAL STRUCTURE

Up to this point all investments were assumed to have been financed with equity, hence the preoccupation with asset valuation models capable of providing accurate estimates of the returns required to invest in risky projects. Investors, of course, have other options, debt being the pre-eminent example. Indeed, since the cost of debt is lower than the cost of equity, owing to the fact that creditors rank higher in the firm's balance sheet than do shareholders, the choice of financing could have a direct bearing on the cost of capital and by extension the firm's capitalisation rate.

In a seminal study, Modigliani and Miller (1958) demonstrated that this intuitively obvious conclusion was wrong, given certain strong assumptions about the nature of financial markets. In a world of perfect capital markets, where no fiscal distortions exist and the risk of bankruptcy is zero, the capital structure of the firm is irrelevant. The market value of the firm and its cost of capital are independent of the way the firm is financed so long as the choice of financing does not affect the probability distribution of the total cash flows of the firm. On this view, the correct capitalisation rate is one that is appropriate to a pure equity cash flow stream in the firm's risk class. Any attempt to create value via leverage would be frustrated, Modigliani and Miller argued, by investors arbitraging prevailing valuation discrepancies.[10]

Critics of Modigliani and Miller's 1958 paper noted that since corporate tax laws virtually everywhere permit the deduction against taxable income of interest paid on debt, but do not offer similar concessions on dividends, leverage creates so-called tax shield benefits. These flaws were remedied in a subsequent paper, where Modigliani and Miller (1963) showed that these benefits are equal to the capitalised value of $(\tau)R$, where τ is the firm's tax rate and R is annual interest expense. By opting to include debt in its capital structure, a firm is able to alter the probability distribution of its cash flows and thus generate benefits for its shareholders equal to the present value of the associated tax savings. By extension, the more debt in the capital structure, the lower will be the corporate tax liability, which in turn results in higher after-tax cash flows thus enhancing the market value of the firm.[11] By this line of reasoning, the firm's optimal capital structure should consist entirely of debt. Modigliani and Miller recognised that as a matter of fact firms do not maximise the amount of debt in their capital structure, in their words, they maintain a 'substantial reserve of untapped borrowing power' attributed to the 'need for preserving flexibility', whatever that may mean.

There are, however, limits to the benefits connected with the use of debt. As leverage increases, so too does the risk of bankruptcy. Creditors will, accordingly, demand compensation for the additional risk associated with increased leverage by charging higher interest rates. As a general matter, the required return to debt will increase non-linearly with increased leverage. But the costs do not stop there: creditors are obviously keen to ensure that firm managers will always act in ways that are consistent with their interests. These expenses, known as agency costs, might be expected like bankruptcy costs to be a non-linear function of the degree of leverage.

How important are these latter costs? Here we shall have to make do with estimates of bankruptcy costs, since there are no data that address agency costs. If, however, we assume not unreasonably that the nature of agency costs approximates to the types of direct costs likely to be incurred in bankruptcy proceedings, then the former should provide a rough quantitative estimate of the importance of the latter. On the other hand, there are potentially significant indirect costs associated with the types of limitations creditors impose upon borrowers to protect their interests including, *inter alia*, restrictions on the kinds of investment projects firms may undertake, minimum working capital requirements and other covenants that constrain managerial decisionmaking (Jensen and Meckling, 1976). All in all, the combined impact both of bankruptcy and agency costs could be quite substantial.

It is interesting to note, therefore, that early studies on the costs of bankruptcy found they were relatively modest. Warner (1977), for example, basing his conclusions on a sample of 11 American railroad companies, estimates that average bankruptcy costs amounted to only $1.9 million, equivalent to between 1 and 2.5 per cent of the market value of the average sized firm. These findings, however, relate solely to the direct costs of bankruptcy; they exclude such indirect costs as lost sales and profits, costs associated with restrictions on firm borrowing and the possibility of having to pay higher managerial salaries to reflect the greater risk of redundancy. Taking these additional items into account would obviously increase the magnitude of bankruptcy costs, but even so do not suggest they are in fact all that formidable.

More recent evidence, by contrast, presents a much grimmer picture. According to the data given in Brealey and Myers (1991) immediately following the announcement in April 1987 of Texaco having filed for bankruptcy, the company's share price fell by $3.375 resulting in the loss of almost 11 per cent of the firm's market value. In addition, the share price of Texaco's largest creditor, Penzoil, also fell and in relative terms by an even larger amount, namely, 16.5 per cent. Thus the combined losses, at $1.5 billion, which Brealey and Myers equate to the market's assessment of both the direct and indirect costs, were equivalent to 12.5 per cent of the combined pre-bankruptcy market value of the two companies.[12]

Putting all the pieces together we are now in a position to explain why firms fail to exploit their potential borrowing capacity fully, despite the obvious advantages of doing so. Following Modigliani and Miller, we can write the market value of a levered firm, V, as being equal to its unlevered value, V_U, plus the present value of the tax shield benefits connected with the use of debt, $V_\tau(D/E)$, where D/E is the debt to equity ratio. Since V_τ is an increasing function of leverage, V will increase linearly with increasing debt. On the other hand, we know that the present value of expected bankruptcy and agency costs, $V_B(D/E)$ and $V_A(D/E)$, respectively, are also an increasing function of leverage, such that the higher is the debt to equity ratio the higher will be such costs. More formally, we can express the value of the firm as

$$V = V_U + V_\tau - (V_B + V_A).$$

The equation highlights the offsetting impact of increasing leverage, which simultaneously raises tax shield benefits and expected agency and bankruptcy costs. Increases in debt will increase the market value of the firm so long as $V_\tau > (V_B + V_A)$. Where $V_\tau = (V_A + V_B)$, there are no additional benefits connected with the use of debt, and the value of the firm is maximised. Any increments to debt beyond that point will result in $(V_A + V_B) > V_\tau$, and so cause the market value of the firm to decline.

More specifically, the firm's optimal capital structure can be defined as that point where higher marginal bankruptcy and agency costs associated with the use of additional debt just offset the marginal tax benefits generated by the resulting higher leverage.

Once we introduce debt into the capital structure, the rate used to discount project cash flows needs to be modified to take account of the tax benefits connected with its use. The revised Modigliani and Miller formulation suggests that after-tax cash flows should be discounted at the appropriate capitalisation rate for an unlevered firm in the relevant risk class plus the incremental benefits connected with the tax deductibility of interest payments. On this view, the weighted average cost of capital (WACC) defines the preferred capitalisation rate for projects. Myers (1974), by contrast, suggests that once we recognise that the choice of financing can be value enhancing, it makes better sense to disaggregate project cash flows into their constituent parts, discounting each at their appropriate rate.

Thus, after-tax operating cash flows should be discounted using an appropriate equity required rate of return, while those cash flows arising from the use of debt should be capitalised at the before-tax cost of debt. The two cash flow streams are then summed to arrive at an adjusted present value (APV), and the decision criterion is, as before, accept all projects where APV > 0. Although around for a long time now, APV has yet to command the same broad approval enjoyed by the weighted average cost of capital. The conceptual superiority of APV as a project selection rule would seem to lie only in those instances where a firm's current investment policy departs radically from past practice; only in this case is APV likely to provide a more accurate financial assessment since only the APV valuation reflects the changed state of company. Otherwise, the WACC provides a perfectly acceptable decision criterion for investments that in all essential respects are equivalent to those in the firm's existing project portfolio (Inselbag and Kaufold, 1997).

THE GULF TAKEOVER AS A CASE STUDY

We conclude this chapter by pulling together the various topics that have been discussed to illustrate how these concepts can be usefully applied. To this end, we shall consider the factors that led to the merger of Gulf Oil Corporation, one of the world's then largest integrated oil companies, with SOCAL, another giant energy company. By way of background we might note that the initial attack against Gulf, by Mesa Petroleum, occurred only months after the first-ever decline in official OPEC prices. Prior to March 1983, when the decline occurred, oil prices had either increased explosively – as in 1973 and again in 1979/80 – or been administered at prevailing market levels through prorationing by leading cartel members, above all, Saudi Arabia.

The costs to OPEC of controlling world oil prices were high: following the second oil shock of 1979, OPEC's share of the world oil market declined, while Saudi Arabia's share contracted at an even faster rate. Thus, the March 1983 reduction in OPEC's reference price, amounting to $4 dollars a barrel or more than 10 per cent of the pre-March level, represented an attempt, first of all, to stem the loss of market share and second to minimise the costs of intra-cartel transfers that made possible OPEC's continued control over the market. The psychological impact of the change was dramatic, signalling the very real possibility of continued downside price risk. It is hardly remarkable, therefore, that energy and energy-related investments were reviewed or that the terms applying to energy financings hardened.

Despite the decline in prices Gulf Oil Corporation continued with its ambitious oil exploration and development programme, costing its shareholders some $2.2 billion annually to maintain the company's hydrocarbon reserves.[13] Mesa's attack on Gulf was, like the other tender offers, based upon the conviction that the company was pursuing the wrong strategy, penalising rather than promoting shareholder interests. At the end of the process, the buying premium on Gulf's shares exceeded 100 per cent. In effect, SOCAL's evaluation of the company's current investment policy was that it was destroying shareholder value equal to the prevailing market price, on a per share basis.

The critical elements that enter into the valuation of Gulf's Exploration and Production (E&P) programme are obviously the price of oil and the rate at which prices were likely to escalate in the future; the cost of capital to be used for deriving the present value of the investment; the average length of time from cash outlay to cash recovery; the value of tax benefits connected with exploration and development; the present value of the oil produced as a result of Gulf's investment programme; and the expected rate of inflation.

The price of oil used to value future cash flows was perhaps the most contentious parameter. Up to March 1983, investment projects were evaluated at the prevailing market price escalated at the assumed US rate of inflation, generally speaking around 5 per cent per annum.[14] In the immediate aftermath of the price decline, the benchmark rate was lowered to reflect the lower OPEC reference price but investors continued to escalate prices at the previously expected rate of inflation. All other things being constant, the decline in oil prices implied a proportional decline in cash flow potential of around 17.5 per cent. If, as expected, the price decline led to widespread industry investment cutbacks, then there could be some cost savings associated with lower drilling and other development expenses thus blunting somewhat the immediate negative cash flow impact.

Gulf's failure to institute a radical reconsideration of its investment programme in the light of recent market developments implies one of several possibilities. The first is that the company may have expected the price decline to be temporary and that prices would in due course revert to or surpass their previous peak. Also, Gulf's investment prospects may have been low cost implying that project economics could withstand a permanently lower oil price. Or, finally, it may have taken the long-term view that it was impossible for an oil company to continue operations if it lacked an adequate hydrocarbon base and was therefore obliged to reconstitute its oil and gas reserves regardless of the cost.[15] The truth is we simply do not know what price assumptions underpinned Gulf's investment programme; what we do know is they were fundamentally out of line with the market benchmark.

The other financial parameters are derived from Gulf's operating results, its income statement and balance sheet. At the start of the takeover process, Gulf had 165.3 million shares outstanding with a market value of $39 per share, making for total equity of $6.4 billion. Gulf also had outstanding debt in the amount of $2.3 billion; thus its total capitalisation amounted to nearly $9 billion, of which just over a quarter consisted of debt. Applying a CAPM valuation, the market premium appropriate to Gulf's equity equalled 10 per cent for a total cost of capital of 21 per cent.[16] Finally, the prevailing rate on AA rated debt, appropriate to Gulf, was then 13.5 per cent; assuming a marginal tax rate of 50 per cent, the after-tax cost of Gulf's debt amounted to 6.75 per cent implying a weighted average cost of capital of 17.3 per cent.

In 1983, Gulf's reserve to production ratio amounted to eight years, implying that a barrel of oil discovered today would be produced eight years later, thus defining the time lag to recovery. In the two years prior to the takeover battle, Gulf had spent $5.4 billion on exploration and production and added oil and gas reserves of roughly 673 million barrels. Dividing reserve additions into the investment cost yields average finding costs of $7.94 a barrel. Of this total, roughly $2.10, attributable to exploration costs, can be expensed immediately, while the remainder ($5.84) has to be capitalised and expensed in eight years time, that is, when the oil discovered today is actually produced.

Combining all of these data, we can calculate the net present value of the tax benefits associated with Gulf's exploration and development programme. The per barrel tax benefit associated with exploration is simply the product of the actual outlay in the previous year multiplied by the assumed 50 per cent tax rate, in the case to hand $2.10 × 0.5 = $1.05; discounted using Gulf's estimated weighted average cost of capital (17%) the present value of current exploration costs is $0.90. The tax benefit associated with production costs is, of course, smaller since it cannot be claimed until eight years in the future and is derived by multiplying the estimated per barrel production cost by the assumed tax rate and converted to its present value by applying a discount factor equal to $1/(1.17)^8$ or $0.82. The present value of total finding cost is simply the sum of the two numbers, namely, $1.72

Gulf's after-tax profits amounted to roughly $2.2 billion or $7.67 on a per barrel basis; this figure provides a convenient base from which to calculate the value of the operating profit to be obtained from producing a recently discovered barrel of oil in eight years time. Since we are assuming the price per barrel will escalate at the same rate as the assumed rate of inflation, the grossed up operating profit is equal to $7.67(1.05)^8$, which when converted to a present value basis is equal to $3.25. Given all of these assumptions, the net present value on a per barrel basis of one year of Gulf's exploration and production programme amounted to:

PV operating profit:	$3.25
PV of E&P tax benefits:	$1.72
PV of finding costs:	($7.94)
NPV for one year:	($2.97).

In 1982 and 1983 Gulf reported average reserve additions of 336 million barrels, meaning that the annual cost of the programme generated a loss of nearly $1 billion. If Gulf had persisted with this policy in perpetuity, the total loss incurred by shareholders would have amounted to $8.3 billion, equivalent to $50.31 per share. If we add the benefit to be derived from immediate suspension of the investment programme to the prevailing share price, the resulting figure ($89) comes remarkably close to SOCAL's winning bid of $80. SOCAL was thus able to acquire reserves at a fraction of its own finding cost; adding in the financial benefits connected with rationalisation, above all, the fact that upwards of 10 000 jobs were eliminated following the merger, resulted in an annual savings equivalent to $7.50 a share.

Three years after the merger, the price of oil collapsed, declining by 36 per cent to $18 a barrel. While it is fair to say that virtually no one then would have forecast the subsequent price collapse, it is equally true that few if any shared the apparent unbounded optimism of Gulf's management, a point proven by the fact that the peak of the industry's investment cycle was reached one year prior to the merger. The only way the investment programme could have made any continuing sense was if some of the key parameters developed far more favourably than those used to generate the base case.

In particular, the results of the financial analysis are obviously sensitive to both the time lag to recovery and the rate of oil price escalation. Holding everything else constant, shortening the recovery lag would have had an immediate positive impact by bringing forward to the present more of the investment's cash flow potential. Suppose for illustrative purposes the time lag to recovery was cut in half, that is, instead of waiting eight years to produce recently discovered reserves they could have been produced within four years. On this basis most, but not all, of the lost value would have been eliminated; in effect, the programme would have made economic sense, with the return just about covering the project's opportunity cost. The same question can be asked of the importance of oil prices; more to the point, what rate of price escalation would have been required for the project to break even in net present value terms? Solving the NPV equation for the rate of price escalation yields an answer of 8.3 per cent.

In the final analysis, we may confidently conclude that the sorts of assumptions used to develop the base case constitute the most reasonable set possible. Gulf may very well have been in the process of reviewing its E&P programme in the immediate aftermath of the March 1983 price decline, but within months of the event its energies were diverted to fending off Mesa's hostile takeover bid. Ironically, the best defence against Mesa's attack would have been to suspend its E&P programme, driving up further the price of its stock and thus raising the cost to Mesa of continuing with the takeover. Indeed, it is possible that is just what Mesa was hoping Gulf would do as it would have resulted in an immediate capital gain on its stake in the company.[17]

There are no indications Gulf was prepared to reconsider its investment programme or pursue other obvious strategies, launching a counter-takeover bid against Mesa, for example, to ensure the company's survival. All we can say for sure is that Gulf's investment programme was conceived at a time when oil prices were far higher, and would just about have made economic sense if OPEC had not lowered its reference price. The destruction of shareholder wealth followed from the company's failure to respond quickly to the change in the state of the oil market. What Gulf failed to do for itself, was eventually forced on it by others.

Notes

1 In the example to hand, the interest payment is received annually. If interest payable on the investment is compounded m times per year, the above expression becomes $FV = PV(1 + r/m)^{mn}$. As m approaches infinity, meaning that interest is being continuously compounded, $(1 + r/m)^{mn}$ becomes $FV = PVe^{rn}$, where e is the base of natural logarithms.

2 In discounted cash flow analysis the initial investment, C, is shown with a minus sign to signify that it is a cash outflow. It is common to treat the initial investment as if it occurred instantaneously, in period $t = 0$; by common convention, the initial expenditure is extracted from the cash flow stream and shown by itself as $-C$. The summation then begins with C_1, when the first positive first cash flow generated by the investment is received. Of course, investment outlays can be, and for major projects, are spread out over a number of years, in which case, each outlay is discounted in the period in which it is incurred. Thus the calculation relates the cash flow generated by the investment

to the cost of the investment. The difference, as we shall soon see, is called the project's net present value. Finally, in the continuous compounding case we can express $PV = FVe^{-rn}$.

3 These results can be compared with those reported in Higson (1991). Despite the fact the CBI survey was undertaken almost ten years after the latest survey given in Higson, the results are virtually identical. More to the point, in both surveys the simple payback rule emerges as far and away the favoured investment evaluation technique, 79 per cent in the survey reported by Higson versus 75 per cent as reported by the CBI.

4 One of these studies looks at the behaviour of investment spending by oil companies on non-oil related activities in the aftermath of the price collapse of the mid-1980s. The evidence suggests that oil companies cut back on their non-oil activities by more than did other firms engaged in the same sectors. The decline in cash flow associated with the oil price decline resulted in a 10–20 per cent decrease in investment spending by energy companies on non-oil-related activities.

5 In general, empirical tests of Tobin's q-ratio as an explanation of the level of investment spending fare badly; see, for example, Abel and Blanchard (1986) and Kopcke (1993). As we shall see, part of the problem appears attributable to the data series that have been used to test the theory; alternatively, the hypothesis may be conceptually flawed (see below), and by extension so too will be the traditional NPV analysis.

6 In terms of the oil exploration example given above, while σ_j, the standard deviation of the return on an oil exploration investment, is likely to be high, the fact that ρ_{jM} is low, means that the investment is subject to little systematic risk and will be priced accordingly. In fact, betas applying to oil companies are fairly low: for US exploration and production companies, the industry beta is 0.92; for integrated majors, beta is even lower. On the other hand, US refining and marketing companies are exposed to far greater risk as shown by an industry beta of 1.22 (Merrill Lynch, 1997).

7 Malkiel and Xu suggest one of the reasons markets may price unique risk is due to portfolio managers requiring an additional risk premium on individual stocks that are perceived to be exposed to extraordinary specific risk. The premium relates to the consequences for institutional investors of holding on to an asset whose price has fallen sharply over some reporting period.

8 Worth noting in this connection is Lawden's (1968) observation that 'Pygmalionlike, man creates his models of the universe and then, succumbing to their fascination, cannot find the heart to destroy them when it is found they are inadequate for their purpose of explaining his experience of nature... This doctrine of the primacy of the theoretical model over the physical phenomena received its full expression in the works of Plato and, after two thousand years, its siren song is still almost irresistible.'

9 Bower *et al.* (1998) compare required returns using the CAPM and APT, and show that the two approaches can produce quite different results. For 12 of the 17 firms included in their sample the authors show APT produced lower required returns than did CAPM. The mean sample return using the CAPM was 23 per cent compared with 18.8 per cent for APT. In 11 cases the two methodologies produced differences at least as large as 4 per cent, a difference big enough the authors assert 'to suggest that companies should use caution in setting screening rates'. See also Groenwold and Fraser (1997) who, basing their analysis on the Australian economy, report similar findings. More to the point, they find the factor set accords broadly with prior research, with financial

factors (inflation and interest rates, for example) more important than either real or economic-openness variables.

10 Imagine two firms in the same risk class, the only difference being Firm A is all equity (E) financed while Firm B includes both debt (D) and equity in its capital structure; both firms have identical earnings of x each year. Now, consider two investments, the first consisting of a 10 per cent stake in the equity of Firm A, which pays 0.1x; the second consists of a 10 per cent investment in the debt and equity of Firm B, with a pay-off equal to $0.1(x - r) + 0.1r = 0.1x$. In efficient financial markets, investments having the same risk–return characteristics should be priced identically: $0.1E_A = 0.1E_B + 0.1D_B$, $E_A = E_B + D_B$ so the value of Firm A must equal the value of Firm B. Capital structure is irrelevant for firm value.

11 Firms pay interest before they pay taxes. Assume the corporate tax rate is τ. An all equity financed firm with earnings of x would pay tax of τx, leaving $(1 - \tau)$ for distribution to shareholders. A leveraged firm with the same annual earnings pays interest of r and tax of $\tau(x - r)$ leaving $(1 - \tau)(x - r)$ available for shareholders. The leveraged firm pays tax of $\tau(x - r)$, compared with τx for the all equity financed firm, thus generating an annual tax savings of τr.

12 This estimate is actually below the industrial sample analysed by Altman (1984), who found that median direct and indirect bankruptcy costs at the time of filing amounted to around 16 per cent of firm value; for retailers combined costs were rather lower, of the order of 8–12 per cent of firm value.

13 Strictly speaking Gulf did reduce exploration expenditure over the course of 1983, by some $400 million compared with 1982. Even so, exploration outlays remained high by historic standards. Even more compelling is that Gulf was using its massive cash flow to buyback its own shares, having initiated the purchases in 1981. The process ended in March 1983, coincident with the decline in oil prices. It is arguable whether the buyback added shareholder value; Gulf, accordingly, would have been well advised to terminate both the buyback and exploration programmes.

14 McKechnie (1983a) presents ten oil price forecasts through to 2000 as seen from 1982. The range of forecasts to the end of the century was quite high: the lowest being $42 per barrel, while the highest was $88. Note that even the lowest projection envisaged an increase of $10 a barrel by 2000, notwithstanding clear evidence at the time these projections were compiled of growing oversupply and its attendant depressing impact on market prices. Only one other projection was below $60, the remainder falling somewhere within a $70–$90 range.

15 The view will resonate with many in the energy industry: the upstream sector exists, they argue, to find and develop hydrocarbon resources, implying that the economic merits of a project are of secondary importance. The issue, as we shall see in the next chapter, is not whether but when to produce an undeveloped reserve. The analysis will show that even if it is uneconomic given prevailing market conditions to develop the resource immediately, the reserve still has, what is called in options theory, time value.

16 This figure compares with estimated required rates of return for international petroleum projects of 13–25 per cent. These estimates apply to projects initiated in the early 1980s, and were converted to nominal returns by applying a 5 per cent inflation rate to the real returns given in McKechnie (1983).

17 By August 1983, Mesa had acquired 9 per cent of Gulf's shares; six months later Mesa made a partial tender that would have increased its interest to 21.3 per cent. At this point, Gulf decided to sell the company to the stronger merger partner.

3 Project Evaluation in Practice

Four basic steps define the practice of capital budgeting. The first and most obvious is development of a comprehensive model, based upon the best available data, that accurately describes the operational and financial structure of the investment. The model is then used to develop a consistent and realistic assessment of the project's cash flow potential. The second critical step is determination of an appropriate rate of return to measure the project's true opportunity cost to the equity investors; in addition, an accurate assessment of the project's debt capacity as well as its potential cost is required. These two parameters jointly define the relevant capitalisation rate for project cash flows, the third critical element in investment appraisal. The fourth basic step is application of sensitivity analysis to identify those project risks that are most critical to the investment's success. Once properly quantified, investors can investigate the cost effectiveness of available options for mitigating or eliminating undesirable project risks.

Many of these issues have already been addressed although in some instances the basic concepts will have to be adapted to take account of the changed context in which they are to be applied. One of the more crucial issues is determination of how to evaluate and price the risks connected with investing in offshore projects, a topic that is covered later in this chapter. Quantification of specific project risks and their mitigation is a second important area, aspects of which are addressed here while others are deferred and considered within the framework of project financing.

Finally, one of the major weaknesses of traditional methods of project analysis is that they tend to be too static. Investments are treated as if they are passively dictated by technical considerations, with the corollary that there are no benefits to be gained from actively managing project assets. Project outcomes, in other words, are expected to evolve more or less in keeping with the base case analysis. There is also an additional implicit assumption that investment decisions are reversible; investors somehow are able recover the costs of failed projects. This sort of thinking has led several analysts to condemn traditional discounted cash flow methodologies for failing to recognise the operational flexibility inherent in most projects, and so provide a poor foundation upon which to base capital budgeting decisions (Hayes and Garvin, 1982; Mason and Merton, 1985; Dixit and Pindyck, 1994; Leslie and Michaels, 1997).[1] On this view, project structures are as, perhaps more, important than is the underlying project.

Plainly, there is much to be said for such criticism. The dichotomous nature of decison making under traditional investment analysis – accept the project if its NPV > 0, reject it otherwise – simply cannot do justice to the complexity of the actual investment process. To cite just one example: the use of current evaluation techniques assumes that acquisition of undeveloped reserves by energy or mining companies is jointly determined with the decision to develop them, when in fact the two decisions are independent.[2] We know from options theory that there is value in acquiring such reserves even if prevailing market conditions do not justify their immediate exploitation. Traditional methodologies, by contrast, would reject the initial investment out of hand. The importance of timing, one among many important issues capable of being addressed by options models, simply cannot be accommodated within the framework of standard capital budgeting techniques. On the other hand, it is too facile to dismiss traditional methodologies. Not all investments lend themselves to the use of the options-theoretic approach, nor is it clear that even when they do, such methods necessarily produce superior results.

DEVELOPING REALISTIC ESTIMATES OF KEY PROJECT PARAMETERS

The development of realistic estimates of major project variables depends upon several broad considerations. Where the relevant parameters are technical in nature, operating or cost factors, for example, and the underlying technology is known, it is relatively straightforward to develop reasonably precise performance estimates. The large number of existing similar projects means it should be possible to construct probability distributions covering most, if not all, of the relevant technical parameters. Where, by contrast, the assessment depends upon market forecasts of, say, output or input prices, it is far more difficult to achieve a satisfactory standard of accuracy. These problems are particularly acute for natural resource projects. The almost universal failure to develop even remotely accurate forecasts of the medium and longer-term outlook for energy and metals prices proves the point.

The significance of this failure is underscored by the fact that prices tend to fluctuate more widely than does production, meaning that the behaviour of output prices is likely to be the more important source of volatility in project revenues and, by extension, project cash flow. This problem is common to all mineral industries, but even more so for petroleum investments, where fundamental changes in the structure of the market – owing partly to the breakdown of the OPEC cartel and partly to sweeping regulatory changes, especially in the natural gas industry – compound the increase in price volatility observable since the mid-1980s. The importance

Chart 3.1   *Crude oil price forecasts, 1983–91*

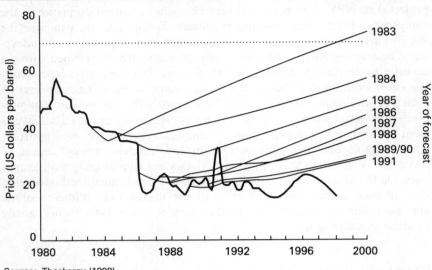

Source: Thackeray (1998).

of forecast accuracy, or the lack thereof, is best illustrated by tracing the
evolution of the price premises used by the petroleum industry to evaluate
project economics. Chart 3.1 shows the consensus outlook for oil prices
since the early 1980s (see also Adelman, 1993); similar figures for the US
natural gas industry are given by Tussing (1996), whose data set consists of
forecasts compiled by the Energy Information Agency (US Department
of Energy).

The chart in effect replicates the base price case that would have been
used by most industry analysts, including major energy banks, for evaluating a given project's cash flow potential at different dates over the past
decade. The sequence of forecasts appears to confirm that historic – and, as
we shall see, current – oil price projections derive from two basic but flawed
assumptions. The first is that the current market price equates to the
petroleum industry's long run competitive price plus a forecast error.
Downward revisions signify that the price was initially assessed as being
too high, with the error corrected in subsequent forecasts. The second key
assumption is that the trend in the long run competitive price is inexorably
upwards owing to inherent scarcity.[3]

The immediate impact of the 1983 decline in OPEC's official reference
price was to rebenchmark the initial price only, generally by an amount equal
to the lower market price. Otherwise, prices continued to be escalated at
previously assumed rates, while natural gas prices, owing to the imminence

of decontrol, were increasingly evaluated at thermal equivalence. Over the following two years not only was the initial price consistently lowered, in keeping with the general decline in market prices that occurred over that period, but the tendency to escalate prices also ceased, at least among commercial lenders (Pollio, 1986). By the third quarter of 1985, project cash flows were evaluated at an unchanged $28 a barrel over project life.

Finally, the chart shows that following the 1986 price collapse nothing changed in terms of the way forecasters viewed the medium- and longer-term outlook for oil prices. Current lower prices again simply corrected past errors, while the possibility that recent market developments might have fundamentally altered the longer-term price trend appears to have been rejected. The only concession in that direction is that in the post-1986 period the predicted price path is flatter than the trend indicated in pre-collapse projections. Nor do more recent oil price forecasts show any departure from past attitudes. A tabulation compiled by the World Bank (1994) confirms that the majority of recent forecasts are for oil prices to rise in nominal and, by the end of the decade, in real terms as well.

The preceding discussion highlights the difficulties encountered in developing baseline estimates of key project parameters; in the case of mineral price forecasts, these problems are compounded, as we have seen, by the consistent misapplication of intertemporal resource pricing models that leads to exaggerated estimates of prices in the out years. Since project cash flows are evaluated in present value terms, the higher the discount rate, the smaller will be the impact of even large forecast errors on project economics. The logic of this view is that long-term forecasts do not matter all that much, since discounting penalises overly optimistic assessments of key project parameters. All that really counts is the ability to generate a reasonably accurate estimate of near-term financial performance which, under most circumstances, will account for the bulk of a project's present value. There is plainly an element of truth in this view and it is possible to accept it without conceding its corollary that modelling long-term project performance simply does not matter. In truth, small differences in the main project parameters can have a major bearing on project economics and accordingly on the decision to proceed with the investment.

Sensitivity analysis is one way of getting to grips with such issues and is used to evaluate the impact that alternative states of specific project variables are likely to have on project economics. Sensitivity analysis provides no decision rules, only insights into the range of outcomes for individual project variables that can convert a positive into a negative NPV. In short, it helps to define which project parameters are crucial for project success. To illustrate the application of sensitivity analysis, we shall revert to our previous discussion of Gulf Oil Corporation's exploration and development programme in the two years prior to its acquisition by Chevron (previously SOCAL).

The aim is not to replicate the original analysis, but rather to illustrate what impact, if any, changes in key project variables might have had on the economics of the investment programme. Nor shall we consider the importance of price sensitivities. The original analysis concluded that the implied price premises were more or less consistent with then-prevailing views of medium- and long-term price trends, a conclusion reinforced by our subsequent discussion of historic oil price projections. In any event, we noted in the original analysis that, all things constant, the investment programme would have been economic, in the sense of producing a zero NPV, if the rate of price escalation exceeded 8 per cent per annum. Given then-prevailing inflation forecasts, the indicated rate of escalation would have implied an annual real rate of oil price increase of more than 3 per cent. No one at the time envisaged prices rising anywhere near that rate; indeed, the balance of near-term risks then were, as we have seen, all on the downside.

There are, however, two other critical project variables – the time lag to cash recovery and the ultimate size of reserve additions – which, unlike the rate of oil price escalation, could have been evaluated using less conservative assumptions. Note, too, that changing the ultimate size of reserve additions would have had a positive bearing on finding costs, given the way these estimates are derived. Consider first the time lag to recovery. In the original analysis, the recovery lag was estimated assuming that Gulf's then reserve to annual production ratio provided a suitable proxy. On this basis, the recovery lag was estimated at eight years. A more plausible figure would have been of the order of three years, typical of the development lag of an offshore US reserve (Siegel *et al.*, 1987). For sensitivity analysis purposes, we shall consider the impact of reducing the recovery lag to three or five years.

The second critical variable, the ultimate size of reserve additions, was initially assessed by utilising actual reserve additions in the two years prior to the takeover. Several objections can be raised against the use of this measure. First of all, it does not relate uniquely to the impact of recent exploration and development expenditure but rather to the cumulative impact of past and more recent investment outlays. The conceptually relevant indicator is, of course, reserve additions generated by the then-current investment programme only. Such figures were unknown and unknowable at the time, hence the necessity for using past reserve additions as a proxy for prospective additions. Nor is the assumption unreasonable: Gulf's latest investment programme was of roughly the same scale as in the recent past, creating the strong presumption that current results would not differ all that much from previous investments.

The second, and far more serious, difficulty is that proved reserves, which is what the bulk of reported additions consisted of, provide an extremely conservative estimate of ultimate reserve additions. By definition, proved reserves consist of those that are capable of being produced with reasonable certainty under current economic and operating conditions (Brown, 1990). However, we know that ultimate recovery is far larger; the figures given by Odell and Rosing (1983) suggest that reserve appreciation can be in the order of three times the level reported initially. These gains are due to a better understanding of the true reserve potential of a given field development based upon actual production experience, but may also require additional investment expenditure to realise higher recovery rates. In the following analysis, we assume that ultimate reserve additions would have exceeded those initially reported by a factor of either two or three, while the additional investment cost needed to achieve higher recovery would have increased development costs by between 50 and 75 per cent. To simplify the analysis, these costs are assumed to have been incurred immediately when in fact the needed expenditure would have occurred much later, although precisely when is a matter of conjecture. The practical effect of this assumption is to overstate costs and hence understate potential economic benefits.

Table 3.1 summarises the impact each of these changes would have had on the value of Gulf's investment programme measured in net present value terms. The results are fully in keeping with *a priori* expectations: shortening the lag to cash recovery, all things constant, leads to an improvement in project returns as does increasing the magnitude of reserve additions. Higher development expenses, by contrast, have a detrimental impact on project economics. Even so, the results suggest that the investment programme's financial benefits were marginal under virtually any circumstances. Only at the extremes of the analysis do the sensitivities convert a negative to a positive NPV. Thus, shortening the recovery lag by

Table 3.1 Sensitivity analysis of NPV to changes in time lag to recovery, reserve additions and development costs (NPV per barrel)

Base case	Time lag to cash recovery		
Eight years	Three years	Four years	Five years
($2.97)	$0.32	($0.50)	($1.24)
		Reserve additions	
	2.0 × base case	2.5 × base case	3.0 × base case
	A. Development costs increased by 50%		
($2.97)	($1.16)	($0.28)	$0.44
	B. Development costs increased by 75%		
($2.97)	($1.79)	($0.78)	$0.04

five years produces a positive per barrel NPV compared with the base case; so too does trebling the scale of reserve additions, a result that holds whether development costs are increased by 50 per cent or at the higher rate. Still, the indicated per barrel NPVs are modest, and in all other cases remain negative. Nothing in the sensitivity analysis, accordingly, would lead us to reconsider our original conclusion that by continuing with its exploration and development programme Gulf was actually destroying shareholder value.

PROJECTS AS OPTIONS

In light of the above analysis the question naturally arises whether an oil company should have invested in acquiring undeveloped reserves following the March 1983 decline in OPEC's reference price and even more so in the aftermath of the 1986 price collapse. Our analysis of Gulf's exploration and development programme led to the conclusion that, under almost any state of the oil market, it would have made better sense for the company to have utilised its cash flow in other, more productive ways. This conclusion was based upon an assessment of the representative project, when in fact Gulf, as any other energy company, would have had a portfolio of projects each with its own unique cost and production characteristics. Using traditional discounted cash flow methods, only those capable of exceeding the risk-adjusted rate of return would have been worthwhile undertaking. Of course, the concurrent decline in drilling and operating costs and changes in tax rates applying to oil developments, especially after 1986, would have altered the economic characteristics of many previously marginal projects. In light of these changed circumstances some of these, too, might have qualified for commercial development.

Still, this hardly appears to be an accurate description of the investment behaviour of the world petroleum industry, which routinely pays out large sums of money for the right to explore for and develop oil and natural gas reserves even in high cost frontier areas. 'If natural resource companies were somehow committed to produce all resources discovered, then they would never explore in areas where the estimated development and extraction costs exceed the expected future price at which the resource could be sold. However, because they can choose when to initiate such development, it may pay to explore in high production cost areas in order to gain the option to produce if the price of the resource at some later date is higher than was expected' (Mason and Merton, 1985). Petroleum companies are in effect acquiring the physical counterpart to financial options – the right but not the obligation to call potential reserve additions.

Given the close mapping that exists between the two, the techniques of the latter can be used to value the pay-off to the former. The potential to acquire future reserves is a multi-stage process that involves the valuation of a set of nested options (Paddock *et al.*, 1988). The first stage of the process is exploration where the firm acquires, through the payment of a bonus or licence fee, the option to spend expected exploration costs and receive the expected value of undeveloped reserves. The exploration programme produces evidence on the quantity, quality and cost of developing the reserve. If positive, the licence holder acquires an option to pay the development cost and obtain in return developed reserves. The final option confers the right to extract the resource.[4]

Like their financial counterparts, each of these options has value in its own right. In options theory, this value derives from two sources: the first is intrinsic value, which compares the exercise price of the option with the current market value of the underlying asset. As with a call option – the analogue to an oil company's right to acquire undeveloped reserves through exploration and development – if the market price of the underlying asset exceeds the price at which the option holder can purchase the asset, it may make sense to exercise the option now or alternatively defer the decision against the expectation that the value of the underlying asset will continue to appreciate. By contrast, if the reverse relationship applies, then the holder will wait until such time as it makes sense to exercise the option; if over the holding period, it never does, the option will expire worthless. In which case, the investor's loss will be equal to the premium paid to acquire the option.

Even if the option currently has no intrinsic value, it still has time value which derives from two sources. One source of time value is leverage: the longer an investor refrains from paying out the exercise price (X), the longer is the investor able to earn interest on the exercise price. The second source of time value is the so-called option feature, the value of not having to buy the underlying asset at the strike price if the market price at expiration turns out to be lower than that.[5] The value of the option feature thus depends upon the probability that the price of the underlying asset will in fact fall short of the strike price at expiration. The probability of that being the case is in turn a function of three variables; (1) how much in- or out-of-the-money the call is, (2) the amount of time remaining for the price to change prior to expiration, and (3) how much of the underlying asset's price usually changes over a given time period, namely, its volatility. More formally, the value of a call option can be written as:

$$C = \max(0, S - X) + [X - PV(X)] + \text{option feature}$$

or,

$$C = S - PV(X) + \text{option feature}.$$

The value of the call (C) thus depends positively upon the market price of the underlying asset (S), the time to maturity, the variance of the price of the underlying asset and the risk-free rate of interest and negatively on the exercise price. The first source of option value derives from the fact that as the price of the underlying asset increases, the higher will be the expected pay-off to the call. The contribution of time to maturity to the value of the option can be thought of as emanating from two sources: (1) the longer the time to expiration, the lower the present value of the expenditure of the exercise price, or equivalently, the higher the return to an investment equal in value to the exercise price – hence the positive relationship *vis-à-vis* the risk-free rate; and (2) the longer the maturity of the option, the greater the likelihood that the price of the underlying asset will rise enabling the holder to exercise the option profitably.

If the volatility of the price of the underlying asset is high, then so too is the probability that the price of the underlying asset will rise, thus increasing the option's pay-off. And finally, the higher the exercise price, the less likely it is that the option will produce a positive pay-off. Standard option valuation models usually assume the underlying asset pays no dividends. If this assumption is relaxed, it is reasonable to suppose that, other things being equal, the higher the dividend yield, the lower the value of the call option, owing to the foregone dividend income.

Undeveloped oil reserves share all the essential characteristics of options and accordingly can be valued in keeping with the same principles that govern the price of a call. Table 3.2, adapted from Siegel *et al.* (1987), clearly illustrates the close correspondence between the two and defines the variables that are required to determine the option's value. Since existing reserves tend to be widely traded it ought to be possible to obtain a direct estimate of both the current value and the price volatility of developed reserves, the first two variables listed in the table. In practice, however, there is no organised exchange on which these assets trade and in many instances oil companies are reluctant to disclose the prices actually paid for reserves acquired this way. These practical limitations can be overcome by following

Table 3.2   Comparison of variables for pricing models of stock call options and undeveloped petroleum reserves

Stock call option	Undeveloped reserve
Current share price	Current value of a developed reserve
Variance of rate of return on share	Variance in the rate of change of the value of a developed resource
Exercise price	Development cost
Time to expiration	Relinquishment requirement
Risk-free rate of interest	Risk-free rate of interest
Dividend	Net production revenue less depletion

Source: Adapted from Siegel *et al.* (1987).

the industry rule of thumb which says that the per barrel value of developed oil reserves is equal to roughly one-third the prevailing market price.

The volatility of reserve values is likewise unobservable, although it seems reasonable to suppose that the standard deviation of oil prices provides a close proxy. The relinquishment rate, corresponding to the time to expiration, is a parameter that depends uniquely on the terms and conditions stipulated in the lease agreement for a particular tract; for contracts covering the US Gulf coast, operators typically have between five and ten years to initiate exploration or development or else forfeit their rights.

The payout rate, the counterpart to the dividend yield in a financial options framework, can be defined as net production revenue less depletion expressed as a fraction of reserve value. Siegel *et al.* (1987) derive its value in keeping with the following set of assumptions. Both the per barrel value of a developed reserve and unit production costs are modelled as a constant fraction, one-third and 30 per cent, respectively, of the crude oil market price while the corporate tax rate net of depreciation allowances is assumed equal to 34 per cent. These assumptions imply that the after-tax profit on a barrel of crude oil is 46 per cent of the market price. Using these values, Siegel *et al.* estimate the payout ratio as just over 4 per cent, which is equivalent to a stock paying the same dividend yield.

Applying these parameters, the option pricing model can be solved to yield an estimate of the value of an undeveloped reserve; the resulting figures, which correspond to alternative reserve value ratios and estimates of the volatility of underlying reserve values, are summarised in Table 3.3. The present value of a developed reserve (V') is determined as $V' = e^{-PL}V$, where V is the market value of a developed reserve, P is the payout ratio and L is the time lag to development. For $V = \$12$, $P = 0.04$ and $L = $ three years, the values used for the numerical example given in Siegel *et al.*, $V' = \$10.61$. The present value of a developed reserve with a current market value of $12 which takes three years to develop and has a payout rate of 4 per cent is $10.61. The final parameter value is obtained by dividing the reserve value into its development cost (D), namely, V'/D. The interpretation of this ratio is identical to the NPV rule: if $V'/D > 1$ it is worth undertaking development of the reserve, otherwise the investment should be deferred or abandoned.

The figures given in the table show the value of an option to develop an undeveloped petroleum reserve, whether the option is currently in the money, that is, $V'/D > 1$, or not.

Consider, for example, the case where $V'/D = 0.9$: while it would be uneconomic now to initiate development, the possibility of being able to do so given a change in market conditions means that the option has time value. At the lower volatility estimate and with five years to relinquishment, the right to develop an undeveloped reserve with a total development cost of DN, where N is the size of the reserve in millions of barrels, would be worth

Table 3.3   Option values per dollar of development cost

V'/D	$\sigma = 14.2$			$\sigma = 25$	
	T = 5	T = 10	T = 15	T = 5	T = 10
Out of the money					
0.70	.00655	.01322	.01704	.04481	.07079
0.75	.01125	.01966	.02410	.05831	.08650
0.80	.01810	.02812	.03309	.07394	.10392
0.85	.02761	.03894	.04430	.09174	.12305
0.90	.04024	.05245	.05803	.11169	.14390
0.95	.05643	.06899	.07458	.13380	.16646
In the money					
1.00	.07661	.08890	.09431	.15804	.19071
1.05	.10116	.11253	.11754	.18438	.21664
1.10	.13042	.14025	.14464	.21278	.24424
1.15	.16472	.17242	.17599	.24321	.27349

Source: Siegel *et al.* (1987).

0.04024DN. With D = $10 and N = 100 million barrels, the total development cost is $1 billion; the right to develop the reserve sometime in the future is worth $40 million today.

Table 3.3 also shows that with volatility given, the longer the time to maturity, the more valuable the option is. Conversely, with the time to maturity given, the option is more valuable the higher is the volatility of the price of the underlying asset. This latter consideration means that a change in market conditions is not confined uniquely to an improved outlook for prices or to lower development costs; higher volatility – anathema in an NPV context – is a source of value in an options-theoretic framework.[6] In the example just given, the undeveloped reserve would be worth $112 million at 25 per cent volatility, roughly three times more than its original value. The higher volatility increases the odds of being able to exercise the option profitably and so creates additional value for the option holder.

We might note in this connection that the lower estimate corresponds to actual price volatility from the early 1970s to the mid-1980s, while the higher estimate is more representative of current trading conditions. Verleger (1994), for example, reports price volatility in the early 1990s at between 20–30 per cent; extending these figures through mid-1997, yields a figure of around 25 per cent. Thus, the transition from OPEC-administered to market-sensitive pricing, as occurred following the 1986 price collapse, has been associated with a marked increase in price volatility.

The options framework provides an alternative, some would say superior, perspective on a wide range of capital budgeting issues. Siegel *et al.* (1987) and Paddock *et al.* (1988) provide evidence on the comparative performance of each methodology. Both tests apply to the valuation of offshore US Gulf leases, the benchmark in each case being the bid per tract derived using traditional cash flow techniques.[7]

The first set of bids consists of discounted cash flow estimates developed by the US Geologic Survey (USGS) of the value of the auctioned tracts, while the second are the bid results similarly derived by an oil company that participated in a lease sale held earlier in the same year. Tract values were also assessed by applying an options pricing model to the same data used to construct the discounted cash flow valuations. Given the use of consistent data, any observed differences should be attributable primarily to the different methodologies used. This will almost certainly be the case in respect of the USGS results, as there is no reason to suppose that reported tract valuations are anything other than the figures generated by the discounted cash flow model. Less clear, is whether the oil company bids are similarly uncontaminated, since the pressures to 'correct' the initial quantitative assessments are likely to be far stronger. Finally, each of the resulting valuations is compared with the average bid per tract and the winning bid.

The mean options-derived value for the 21 tracts surveyed by Paddock *et al.* amounts to $4.13, about 16 per cent lower than the USGS discounted cash flow valuation and about a third lower than the mean industry bid. The options and USGS valuations are closely correlated, although both show little correspondence with either average or winning industry bids. The first result is not too surprising, given that the two valuations are based upon the same underlying geological data; that latter finding, however, suggests that industry evaluations differ radically from those compiled by the USGS, although whether this reflects different assessments of geological potential or underlying price assumptions or both is an open question.

Options valuations are also shown to be extremely sensitive to alternative price assumptions. For example, substituting the indicated higher gas price for the thermal equivalent price used in the initial analysis results in an average tract value nearly twice as high as before; however, the gap *vis-à-vis* the average industry bid is about as wide as previously, but now the options valuation is above rather than below the industry mean. The Siegel *et al.* results are similarly equivocal. The mean options valuation over the eight tracts analysed exceeded the average industry bid by a margin of 75 per cent; the company bid, by contrast, was about two-fifths the options valuation, but within 25 per cent of the actual mean industry bid. On the other hand, the options approach more closely approximated to the winning bid, the margin of difference being less than 25 per cent.

It is difficult to draw any firm conclusions from these results, as they are no more (or less) accurate than valuations based upon discounted cash flow techniques. This in itself is a significant finding in the sense that the superiority of such methods is alleged to lie precisely in their ability to capture and value so-called managerial options, which cannot be assessed within the framework of more traditional valuation methodologies. Following Trigeorgis (1993), a project's strategic NPV is equal to the sum of its static NPV plus the value, if any, to be derived from active project management. As long as the second term is positive, a project's strategic NPV must exceed its static NPV. Using this criterion, the Paddock *et al.* (1988) results summarised in Table 3.4 have to be regarded as extremely disappointing in that the static NPV, corresponding to the traditional discounted cash flow analysis used by the USGS, actually produces a higher average tract bid price than does the options valuation model, equivalent in the present context to the strategic valuation. Only when the authors substitute higher natural gas prices for thermal equivalence, the assumption used to derive the base case, do the results conform with those expected.

While it would be wrong to make too much of this finding, it does appear to challenge earlier views to the effect that traditional methodologies are too conservative and accordingly will always undervalue project investments.[8]

Table 3.4   Actual and simulated bids for two US Gulf of Mexico lease sales, 1980 (millions of US dollars)

Valuation methodology	Sample mean	Standard deviation	Coefficient of variation	Ratio to option valuation
November 1980 lease sale (N = 21)				
Option (1)	4.13	5.56	1.35	1.00
Option (2)	8.20	9.42	1.15	1.99
USGS	4.93	6.32	1.28	1.19
Industry:				
Mean bid	6.03	3.58	0.59	1.46
High bid	18.95	16.07	0.85	4.59
September 1980 lease sale (N = 6)				
Option	42.18			1.00
Company bid	17.90			0.42
Mean industry bid	24.33			0.58
High bid	58.54			1.39

Source: Paddock *et al.* (1988); Siegel *et al.* (1987).

Note: Option valuation (1) is based upon $2/mcf gas prices, while valuation (2) is based upon gas prices of $3/mcf.

More recent studies are less dogmatic, arguing instead that the two methodologies, far from being mutually exclusive, can be complementary. Kemna (1993), for example, maintains that the same fundamental principles underpin both discounted cash flow analysis and options pricing models; that the two methodologies can and should be used in tandem when there are future decision points which influence the riskiness of the cash flow; and finally, traditional methods of analysis appear adequate for the evaluation of most expansion projects.[9] The safest conclusion to be drawn from the preceding discussion is where managerial options are important to project success and the investment is irreversible, options valuations would appear to make better sense. Where, by contrast, investments are reversible or lack flexibility with respect to timing, traditional techniques appear to be the more appropriate yardstick against which to measure the project investment.

INTERNATIONAL CAPITAL BUDGETING

Up to this point we have been concerned exclusively with issues arising in connection with the evaluation of domestic projects. In the remainder of this chapter we shall expand the scope of the analysis to encompass international investments and the unique issues they create for project evaluation.

Foreign direct investments are driven by numerous economic and financial considerations; in the case of mineral industries – which up until fairly recently accounted for the bulk of such investments – foremost among these would be geological potential, location and the costs associated with developing and extracting the resource. At bottom, however, the rationale for undertaking an offshore investment, whether within or outside the natural resource sector, must be the same as for a domestic project. Thus the same principles and techniques used to value a local project can be applied, *mutatis mutandis*, to an international investment.

Several analysts, however, point to the possibility that international investments may create diversification benefits that should be considered as an integral part of the investment process. Suppose in this connection a firm is considering two investment options to expand current production capacity. Either it can construct a production facility in its home market or else establish a new subsidiary abroad. Assume further that the prospective returns from the two projects are identical, but the standard deviation of foreign earnings is higher than is the variance of expected domestic cash flows. Imagine, finally, that the cash flows from the local project are more highly correlated with those generated by the firm's existing assets than are the cash flows produced by the proposed offshore affiliate.

The resulting diversification benefits would appear to argue strongly in favour of the offshore investment. With less systematic risk, the required return to the investment will be lower and by extension so too will be the firm's beta. With lower overall business risk, the firm should also be able to access higher levels of debt and on more favourable terms implying a reduction in its weighted average cost of capital. On the other hand, agency costs could be higher for internationally diversified companies than for domestic firms, while multinational companies are exposed to greater political risk.

This stylised description of international investment suggests a number of testable propositions. Agmon and Lessard (1977) long ago evaluated the hypothesis that financial markets reward international diversification. Their test consists of sorting American companies by the proportion of their sales that are generated internationally and beta. If there are benefits associated with international diversification the two variables should be negatively correlated. Their results, although suggestive, are equivocal. For example, American firms with international sales accounting for less than 10 per cent of total turnover had betas of around 1.05; the comparable beta for firms with international sales in excess of 40 per cent is 0.88. On the other hand, firms with international sales of 13–17 per cent actually had lower betas than most companies having a progressively larger foreign sales share. Across the entire sales range tested, the rank correlation between the two variables is −0.56.[10]

The same equivocal findings apply in respect of the other putative benefits connected with international diversification. The hypothesis that, since multinational firms are assumed to have lower business risk than domestic companies, they should have larger debt capacity and accordingly more debt in their capital structure enjoys little empirical support. Burgman (1996), for example, tests both of these propositions basing his analysis on the financial results of nearly 500 US companies, roughly half of which were multi-nationals. With respect to business risk, the results are shown to be sensitive to the particular measure chosen. When proxied by the standard deviation of the change in earnings before interest and taxes (EBIT) and scaled by the level of firm assets, the hypothesis is affirmed; substituting the coefficient of variation of the level of EBIT as a surrogate for business risk, leads to just the opposite conclusion.

The findings concerning debt levels are more straightforward and again are at odds with standard diversification arguments. Between 1964 to 1983, the period covered by Burgman's analysis, domestic companies had average debt levels of just under 33 per cent as against 27 per cent for multinationals; moreover, the null hypothesis that the observed differences could be due to chance is rejected.[11] Nor are these conclusions affected either by firm size or industry-specific effects. Part of the explanation for these negative findings may be that multinational firms are subject to considerably higher agency costs than are domestic companies. Whatever tax shield benefits they may derive from the use of debt are more than counterbalanced by the resulting higher agency costs.[12] Indeed, Burgman's analysis suggests that multinational capital structure decisions may be largely independent of tax shield benefits. Instead, they appear to be motivated largely by foreign exchange and political risk considerations, such that the greater is the firm's exposure to either of these risks, the higher will be the desired degree of leverage.

To summarise, the precise factors affecting foreign direct investment decisions are still only imperfectly understood, but the weight of evidence suggests that diversification may not be among the more important determinants. This finding is fully in keeping with our previous discussion which showed that diversification only makes sense where the firm can achieve these benefits at lower overall cost than shareholders; otherwise, financial markets are more likely to penalise than reward such behaviour.

No such ambiguity surrounds the general financial principles governing international capital budgeting decisions. First of all, it is universally agreed that the only relevant cash flows are those that can be repatriated to the parent corporation. If cash flows are freely repatriatable, the same principles discussed in connection with the evaluation of domestic projects apply equally to international investments. In the case to hand, cash flows are evaluated in local currency terms and converted to the home country currency at expected exchange rates. Once the currency translation has been made, valuation is based upon the standard NPV rule, with the expected cash flows discounted at the investor's opportunity cost of capital. Under traditional valuation methodologies, the discount rate should be adjusted to take account of any benefits (or risks) associated with the foreign investment. We have already shown that financial markets do not appear to impute any value to the potential benefits associated with international diversification. Since project hurdle rates tend to be considerably higher for offshore than for comparable domestic investments, we shall have to look more closely at the factors causing such returns to diverge.

The surveys discussed in Chapter 2 showed that while most firms utilise quantitative methods to evaluate project economics, the vast majority of respondents indicated a preference for conceptually inferior measures, above all, the simple payback rule. The same general conclusions apply in respect of the evaluation by American multinationals of offshore projects (Buckley, 1996a).[13] Even more discouraging is the clear evidence of a regression over the past 30 years towards the use of less sophisticated analytical measures. For example, in the earlier half of Buckley's sample period, the surveys indicate growing use of the weighted average cost of capital as the favoured capitalisation rate. The importance of this measure appears to have declined progressively over the second half of the sample period so that by the early 1990s only around one in five survey respondents reported using the weighted average cost of capital as the preferred yardstick. In the latest survey tabulated, the cost of debt is shown to be the single most important proxy for the project's required rate of return, while the majority of respondents indicated that risk adjustments were typically based upon subjective considerations; shortening the payback period, for example, emerged as the preferred solution for dealing with risk.

It is difficult to know to what to make of these results. All we can say is they are not out of line with what has been observed in other surveys. For example, the CBI survey covering British manufacturing firms discussed in Chapter 2 indicated a clear preference for the same measures favoured by American multinationals, while the principles applied by German banks appear to be based upon even cruder indices (Alexander, 1995).

Taken at face value, the single most important criterion applied by German banks is the profitability and liquidity of the borrower, closely followed by management quality. While leading German banks do appear to take into account, albeit secondarily, the nature of the project they are being asked to finance, there appears to be no concern at all with prospective project returns. In part, this may simply reflect the nature of German lending practices, which favour collateralised over non-collateralised lending. Roughly 80 per cent of the number and 70 per cent of the value of German commercial and industrial loans are collateralised. Even so, the overwhelming impression gained from these results is that the borrower's reputation counts for more than does the investment, an attitude not far removed from the principles governing lending practices in less mature banking markets.

POLITICAL RISK

Exposure to political risk is one of the key distinguishing features of foreign project investments. Political risk may be defined as any action taken by host governments that impairs project economics. Such actions can range from outright expropriation to more subtle forms of confiscation including such things as raising port or rail charges as a disguised means of capturing resource rents, a technique favoured by many mineral-producing countries, including those in the industrial world, Australia, for example (Pollio, 1992). Many of these risks can be mitigated or eliminated through the provision of official or private insurance. The market tends to be fairly broad for many of the more common types of political risks, nationalisation, for example; in other instances capacity is limited so that coverage is slight in relation to the total exposure and the cost accordingly is extremely high. Insurance having to do with guaranteeing financial performance falls largely into this category.

Shapiro (1989) presents an interesting analytical framework for evaluating key political risks, the starting point for which is the recognition that investors forfeit all cash flows subsequent to a political risk event. Political risk events that occur early in project life create greater losses in a present value sense than do those that occur later on; hence the importance international investors attach to the payback rule. On the other hand, most foreign investors will seek to hedge against a broad range of political risks by obtaining insurance against specific events. Indeed, most international loan agreements not only stipulate the existence of such insurance as a precondition for obtaining project financing, but also require an assignment of the proceeds in the event that sponsors have to make a claim under the policy. In the simplest case, we shall assume the investment is financed entirely with equity. In this case, the ultimate financial impact equals the loss of all cash flows subsequent to the political risk event, net of compensation, if any, that might be paid by the host government, insurance proceeds and tax deductions that investors may be able to claim owing to the loss of the investment.

To evaluate the financial consequences of a political risk event, say, expropriation, we need to modify the original NPV calculation to take simultaneous account of the costs and offsets that arise following expropriation. More formally,

$$\text{NPV} = -C_0 + \sum C_i/(1 + r)^i + P/(1 + r)^b.$$

The first two terms on the right-hand side of the equation are as before, namely, the initial investment outlay (C_0) and the sum of cash flows generated by the investment up to the date of expropriation ($b - 1$). The final term of the expression is the present value of any compensation from whatever source (P) received by the investors subsequent to the political risk event, discounted at the investor's cost of capital from year b, when the expropriation occurred. The value of the investment in net present value terms is equal to cumulative pre-expropriation cash flows plus any compensation received by the sponsors owing to the expropriation less the initial cost of the investment. The project will still have produced a positive pay-off to the investors if the financial inflows from all sources equal or exceed the investment cost of the project.

Since all international investments are exposed to political risk, to a greater or lesser extent, it is possible to quantify the probability of the project producing a positive pay-off given a political risk event occurring any time over project life. Following this line of reasoning, we can restate the project acceptance rule as accept all projects having a break-even probability (p_i) in excess of the firm's critical investment threshold, namely, $p_i > p_{Critical}$. This rule appears far superior to the simple payback rule – the criterion apparently favoured by the majority of respondents to capital budgeting practices surveys – in that it takes explicit account of the project's cash flow potential, the risks connected with undertaking an offshore investment and the interactions between the two. The only advantages of addressing political risks by shortening the payback period are ease of implementation and minimal data requirements. However, the strong subjective element here invites the risk of rejecting a project that on other criteria, including the approach advocated by Shapiro, would be acceptable.

The data needed to implement Shapiro's approach are derived from the base case analysis of the project while the value of compensation is in principle known from the terms and conditions applying in the policies covering specific political risk events and from existing tax rules governing expropriations. Less certain is the value of host government compensation both in respect of amount and whether it will be payable in an acceptable form, cash, for example, as opposed to fixed rate bonds. Given this information, Shapiro (1989) derives the following expression for assessing the probability of the project breaking even, that is, producing an NPV equal to zero. All terms are as before with p_b being the probability of expropriation occurring in year b.

$$-C_0 + \sum C_i/(1+r)^i + p_b P_b/(1+r)^b + (1-p)\sum C_i/(1+r)^i$$

If the project is expropriated, then $p = 1$ and the third term disappears meaning that all post-expropriation cash flows are lost. In which case, the value of the project equals the NPV up to year $b-1$, plus any compensation to which the investors are entitled. For present purposes we assume compensation is paid as a lump sum in the year in which expropriation occurs; if for any reason the payment is delayed or paid out over a number of years, the ultimate value of the compensation will be correspondingly reduced. If, by contrast, expropriation does not occur, then $p = 0$, and the second term drops out. The project's NPV now equals the sum of cash flows up to year $(b-1)$ plus all subsequent cash flows generated by the investment from year b onwards.

The probability of the project breaking even can be found by setting the above expression equal to zero and solving for p.

$$p_b\left[\sum C_i/(1+r)^i - P/(1+r)^b\right] = -C_0 + \sum C_i/(1+r)^i$$
$$p_b = \left[\sum C_i/(1+r)^i - C_0\right]/\sum C_i/(1+r)^i - P_b/(1+r)^b$$

The probability of the project breaking even given a political risk event occurring in year b is equal to the net present value of the project through year $(b-1)$ divided into cumulative cash flow through year b less any compensation received by the investment's sponsors. As the expression makes clear the magnitude of p depends upon the level of compensation: any reduction in the scale of such payments automatically lowers the breakeven probability. With suitable modification, the expression can be used to evaluate any political risk event. Buckley (1996a) provides a more general expression for calculating the break-even probability. The base case NPV will always be reduced by the probability of a political risk event occurring multiplied by the present value of the cash flows forgone contingent upon occurrence. Since,

$$NPV = -C_0 + \sum C_i/(1 + r) - p(\text{PV of foregone cash flows}).$$

Setting NPV = 0 and rearranging yields

$$p = -C_0 + \sum C_i/(1 + r)^i/\text{PV of foregone cash flows}.$$

The break-even probability equals the ratio of the project's base case NPV to the cash flows foregone contingent upon any political risk event occurring.

The traditional approach to incorporating political risk in project evalution is to adjust the required equity return for any downside risks; a higher required equity return will cause the weighted average cost of capital to rise and with it the project's overall capitalisation rate. On the other hand, modern finance theory indicates that the required return should take account of any diversification benefits that may arise from undertaking an international investment. Thus the appropriate discount rate for an offshore project is conceptually a function of two possibly countervailing influences that at the margin may offset each other.

To derive a discount rate for an offshore project, the first step is to determine the risk premium to be applied to the cost of equity. Within a CAPM framework, beta determines the magnitude of the risk premium. The beta relevant to an offshore project incorporates both the volatility of the target market to the investor's home market and the covariance of these changes again relative to the investor's home market. In practice, the beta of an offshore project is the product of the beta of a comparable home country project and the country beta (Lessard, 1996); implicit in this formulation is that the project beta applying to an investment in a foreign country is identical to the same project undertaken in the investor's home country.

The degree of economic risk applying across countries, typically proxied by stock market returns, varies with the level of economic development. Moreover, economic fluctuations in emerging markets tend to be largely independent of developments in the OECD countries, signified by the low correlation of stock market returns. For example, using the S&P 500 index as the benchmark portfolio, the correlations *vis-à-vis* Argentina, Mexico and Brazil and the US market are 0.32, 0.22 and 0.40, respectively. Outside the western hemisphere the correlations decline even further. Thus the covariance between the US and Taiwan markets is only 0.09, while the Korean and US markets are actually negatively correlated. There appears to be, in short, little systematic risk inherent in emerging market investments, a point clearly demonstrated by the betas tabulated by Lessard (1996). Abstracting from political risk, the benefits of international diversification appear to be overwhelming.

Of course, this conclusion needs to be tempered by the fact that investments in emerging markets are exposed to varying degrees of political risk, which could offset most of these benefits. Numerous indicators exist that could be used to adjust discount rates to take account of political risk. The most widely used in international capital budgeting are risk premiums applying on international bonds, insurance premiums charged by political risk insurers, export credit agencies, for example; and country and political risk ratings compiled by leading international financial publications, the *Institutional Investor*, *Euromoney* or the Economist Intelligence Unit (see Table 3.5). Each of these indicators has its strengths and drawbacks. Export credit agency premiums are non-comprehensive and thus exclude certain types of risks that could have a devasting impact on project economics; in the important area of infrastructure investments, for example, the failure to concede timely rate increases could result in project failure.[14]

A recent study by Haque *et al.* (1996) provides interesting insights into the economic content of country risk ratings. First of all, the authors show, not surprsingly, that the three indices are highly correlated; pairwise correlations are uniformly high, ranging between 0.8–0.9 or higher for most years, while the Kendall coefficient of concordance, which measures the degree of closeness among three or more variables, is also of the order of 0.9 for the five years tested. They further show that fluctuations in the various indices can be related to a relatively small number of quantitative and qualitative variables that collectively explain much of the variance in the individual ratings. Notwithstanding the high correlation that exists among the indices, the empirical results indicate broad differences in both the magnitude and significance of each index to the underlying set of explanatory variables.

Country risk surveys have long been used as the standard benchmark against which political risk is measured. Such indices are comprehensive and unlike export credit agency premiums tend not to be affected by national commercial considerations. Moreover, the leading surveys are developed using different criteria and combine quantitative analysis with expert opinion. The following table compares the main features of the three leading political risk indicators. Each differs in important ways. The index compiled by the *Institutional Investor* relies upon assessments provided by leading commercial banks, with higher weights attaching to the opinions of banks having larger international exposure. The *Euromoney* index is based upon a relatively small number of economic and financial indicators, with the greatest weight attaching to the subset of analytical and market measures. Finally, the EIU index heavily weights traditional credit indicators, closely followed by political and economic variables.

Table 3.5   Rating agency criteria for assessing country risk

Rating agency	Ratings criteria	
Institutional Investor	Each country graded by 75–100 leading international banks on a scale of 0–100, with 100 representing least chance of default.	
	Individual responses are weighted using a formula that gives greater importance to responses from banks with greater worldwide exposure.	
	Criteria used by individual banks are not specified.	
Euromoney	*Analytical indicators*	*40%*
	Political risk	(15)
	Economic risk	(10)
	Economic indicators[1]	(15)
	Credit indicators	*20*
	Payment record	(15)
	Rescheduling	(5)
	Market indicators	*40*
	Access to bond markets	(15)
	Selldown on short-term paper	(10)
	Access to discount available on forfeiting	(15)
Economist Intelligence Unit	Medium-term lending risk[2]	45%
	Political and economic policy risk	40
	Short-term trade risk	15

Source: Haque *et al.* (1996).

Notes:
1 Debt service/exports, external debt/GDP, balance of payments/GDP.
2 Total external debt/GDP, Total debt service ratio, interest payments ratio, current account/GDP, saving–investment ratio, arrears on international bank loans, recourse to IMF credit, degree of reliance on a single export.

Some idea of these differences is given in Table 3.6, which compares the elasticity of the individual indices with respect to the various variables shown in the right-hand column. The first point worth noting is that all three indices exhibit persistence effects, meaning that individual country scores do not change all that much from year to year. The effect, captured by the lagged dependent variable, is most pronounced in the *Institutional Investor* and Economist Intelligence Unit indices. The *Euromoney* index, on the other hand, exhibits the strongest reponse to the indicated economic measures, and appears far more sensitive to changes in financial market conditions, proxied by the three month US Treasury bill rate, than is true for the two other indices. The *Institutional Investor* index also picks up the interest rate effect, but the magnitude of the elasticity is only about a third as large as the response indicated for the *Euromoney* index. Each index is affected by both structural and location factors, but again the *Euromoney*

Table 3.6 Elasticity of political risk index with respect to selected economic, financial, external and location variables

Variable	Institutional Investor	Euromoney	EIU
Economic variables			
Export growth	0.002		0.016
Current account balance/GDP	0.008	0.014	0.017
International reserves/imports	0.013	0.027	0.044
GDP growth rate	0.021	0.021	
US 3-month Treasury Bill	−0.111	−0.291	
Lagged dependent variable	0.628	0.191	0.761
Qualitative variables			
Primary commodity exporter	−0.011	−0.059	−0.026
Fuel exporter		−0.027	-0.027
Service exporter and recipient of private transfers	−0.006	−0.025	
Diversified export base		−0.020	
Location variables			
Africa			
Asia	0.006	0.023	
MidEast		0.009	
Europe		0.013	
Financial classification			
Diverse sources of borrowing			
Mainly from official sources			
Mainly from market sources		0.022	

Source: Haque *et al.* (1996).

Note: Only statistically significant elasticities are shown in the table.

index shows the broadest overall sensitivity to such variables. Finally, sources of international finance, that is, whether emerging markets obtain the bulk of their external finance via government loans or from the market, help to explain fluctuations in the *Euromoney* index but have no comparable importance for the other indices.

There are two significant practical limitations connected with the use of political risk indices for purposes of adjusting country-specific discount rates. First of all, they are concerned more or less uniquely with credit issues, and thus are of more immediate interest to commercial lenders than project sponsors. For example, they impart no information with respect to expropriation or other political risk events that are of greater importance to direct investors, and thus suffer from the same defects as insurance premiums. It could, of course, be argued that a general deterioration in international creditworthiness provides evidence of the kinds of broad policy shifts that could just as easily have ramifications for direct investors.

On the other hand, there are numerous instances of countries having nationalised foreign assets or other similar hostile acts, without in any way jeopardising their international creditworthiness. The underlying measures used to construct such indices do not take account of such actions and so their value to direct investors, if they have any, lies only in their signalling function. On this basis, the so-called economic freedom indices, which explicitly seek to capture a given country's commitment to a liberal economic regime, might provide far more useful information (Gwartney *et al.*, 1996). Even if they did, investors would still face the same difficulties applying these measures as they do with current country risk ratings, namely, the inability to translate easily or directly individual country scores in ways that permit direct quantitative adjustment of country-specific discount rates.

This latter deficiency goes a long way towards explaining the current strong interest in international bond premiums which do appear to provide the needed quantitative correction. Again, however, they relate more to credit issues, rescheduling or default risks, for example, than to the more conventional concerns of direct investors. Moreover, not all countries have tapped international bond markets, while for many that have trading is so thin that the resulting spreads may not provide an accurate reflection of the country's credit quality. In either case, it appears possible to infer the 'correct' risk rating by focusing on the spreads indicated for other countries that share the same economic and financial characteristics as the country of interest. But even liquid debt is not without its problems: spreads relative to US Treasury securities are extremely volatile, while the securities of countries having a variety of international issues often trade at different spreads. True, these differences may be attributable to differences in the specific features of the individual issues or to technical anomalies in the market. But until these differences are clarified use of such data invites the risk of either penalising or rewarding country-specific projects.

HURDLE RATES FOR OFFSHORE INVESTMENTS

Despite the various difficulties just enumerated, sufficiently varied information appears to exist upon which to base quantitative judgments concerning the importance of political risk factors and how they are likely to affect the valuation of offshore investments. Seen this way, the tendency of investors to adjust project hurdle rates subjectively, as disclosed in capital budgeting surveys, may provide a distorted view of the actual process. If 'subjectively' in this context refers to the way investors translate the wide array of quantitative and qualitative data into a form suitable for adjusting discount rates, they can hardly be faulted. It is less clear whether investors make use of the correct conceptual model; that is, in deriving required returns on offshore projects, to what extent do investors incorporate into their analyses all information relevant to the investment. The following quote, attributable to the chief financial officer of GTE, a major US telecommunications firm, illustrates both the process and limitations of the way international capital budgeting is practised.

> First of all, we calculate different discount rates for different countries... Conceptually we begin with our weighted average cost of capital for domestic telephone investments. Because telephone investments tend to be highly leverageable and because interest on debt is tax deductible, our domestic telephone investments tend to have relatively low costs of capital. But in coming up with an overseas discount rate, we add a country premium to that domestic cost of capital that reflects various country risk ratings as well as the yields on stripped Brady bonds and sovereign debt. We typically try to make some sort of moving average for those market variables to avoid the possibility that spikes in such rates will distort our analysis. The net result is that for most of these countries we end up using discount rates in the high teens or low twenties' (Smith, 1996).

Superficially, the quote provides clear confirmation of the textbook description of international capital budgeting. In most essential respects it confirms the process outlined in earlier sections of this and the previous chapter. Different discount rates are applied to investments in different countries. The basic step in each case is the domestic cost of capital, heavily weighted in this case by debt so as to take full advantage of tax subsidies. Country risk indices together with bond yields are used to determine country risk premiums. Care is taken to smooth the market data so as to avoid introducing any distortions into the analysis. The upshot is that the cost of capital for offshore telephone investments is apparently considerably higher than applied to comparable domestic investments.[15]

It is important to note what is missing from this description. Nothing is said about the appropriate equity return, an omission that perhaps is not too surprising since it might have involved giving away competitive information. The use of the weighted cost of capital is correct to the extent that international projects are financed with the same mix of debt and equity as are purely domestic projects; otherwise, it provides the wrong lens through which to judge an offshore investment. GTE, we are told, fully exploits tax shield benefits available to the company, again fully in keeping with what financial theory predicts; whether the same is true for its offshore investments is unstated. Nor does GTE appear to attach any significance to potential diversification benefits associated with international investments.

The hurdle rate for offshore investments is calculated (implicitly) as the product of the project and (unadjusted) country betas; no allowance appears to have been made for the possible low correlation of cash flows between the target and home countries. Political risk is factored into the analysis, presumably subjectively, via the use of country risk ratings and international bond yields.

This or similar approaches to international capital budgeting seem to result in the application of discount rates far higher than appropriate and thus penalise profitable investment opportunities. Table 3.7 compares the theoretical 'correct' discount rate, the one that is priced solely in keeping with systematic risk, with an alternative rate which focuses on total risk. The country betas shown in the third column of panel A of the table are derived as the ratio of stock market volatility (a proxy for macroeconomic risks), in the individual markets shown, to the US S&P 500. The fourth column lists the correlation between the US market and that of the five emerging economies shown, and confirms the conclusion noted above concerning the unrelatedness of returns between the US and most emerging stock markets. The adjusted country beta is the product of the third and fourth columns and thus corresponds directly to the traditional measure of beta as defined by the CAPM: the only risk priced in the market is systematic risk,

Table 3.7 Comparison of the total risk and CAPM approaches

A. Calculation of alternative betas				
Country	Volatility	Country (project) beta	Correlation with US	Adjusted country (project) beta
Argentina	54.74	5.65	0.32	1.81
Brazil	53.75	5.55	0.40	2.22
India	29.95	3.09	0.03	0.09
Thailand	30.91	3.19	0.23	0.73
Philippines	32.05	3.31	0.22	0.73

B. Calculation of required rates of return				
Country	Credit spread	Risk premium (1)	Risk premium (2)	Cost of equity (1)	Cost of equity (2)
Argentina	4.0	31.08	9.96	28.0	15.97
Brazil	4.1	30.53	12.21	28.4	17.43
India	1.6	17.00	0.51	17.8	7.90
Thailand	0.6	17.55	4.02	17.1	9.01
Philippines	2.0	18.21	4.02	18.9	10.41

Source: Lessard (1996) and Godfrey and Espinosa (1996).

which abstracts from the covariance between a given market and the benchmark portfolio. In most instances, the theoretically correct measure of beta is a fraction of the unadjusted measure.[16]

The second panel of the table shows CAPM-derived estimates of required equity returns on offshore projects in each of the five countries included in the sample. The cost of equity is calculated according to the following expression:

$$r_{equity} = \text{risk-free rate} + \text{credit spread} +$$
$$(0.6)\text{project beta} \times \text{market risk premium}$$

where the risk-free rate is assumed equal to 6.00% and the market risk premium is 5.5%. The project betas are based upon the alternative measures described above, while the home country project beta is assumed equal to one. The impact of political risk is factored into the analysis using international bond premiums; the indicated correction factor is intended to take account of the fact that credit spreads typically explain about 40 per cent of equity volatility in emerging markets (Godfrey and Espinosa, 1996). Allowing for diversification effects has a dramatic impact on required equity returns: the premiums based upon the total risk beta are three to four times larger than those implied by traditional CAPM estimates. Once political risk considerations are taken into account, required returns, depending upon which measure of beta is used, can differ by as much as a factor of two.

Godfrey and Espinosa (1996) argue that required returns based upon the use of uncorrected betas are superior to those derived using more traditional approaches for the following three reasons:

- managers of multinational companies are often interested measuring risk independently of their home portfolio or the world market portfolio. The performance of local managers is likely to be measured using rate of return benchmarks that are comparable to those held out for US managers, that is, benchmarks that give managers little credit for any diversification benefits their operations may bestow upon the company's investors;
- even when management considers the diversification benefits of a project, they probably do not think in terms of diversification relative to the world equity portfolio. Instead they are likely to view it relative to their company's portfolio of productive assets, which in most cases will differ significantly from the global portfolio;
- given heightened uncertainty connected with investments in emerging markets and the practical reality of capital budgeting in large corporations, the use of higher discount rates may better accomplish the aims of capital budgeting systems – to send strategic planners clear signals indicating the full extent of project risks and provide a basis for higher standards of profitability.

Notes

1   Hayes and Garvin (1982) conclude that major investment decisions should be made on the basis of judgement and strategic considerations alone, so as to eliminate the distortions caused by the use of quantitative methods. Mason and Merton (1985) accept that current capital budgeting techniques are flawed but reject the idea that quantitative methods should be scrapped in favour of the subjective and strategic criteria advocated by Hayes and Garvin. The problem, in their view, lies in the shortcomings of particular evaluation techniques not with the quantitative method *per se*. In particular, they argue that option-theoretic models are well suited to addressing the real options inherent in projects. Dixit and Pindyck (1994), by contrast, are even more emphatic about both the shortcomings of traditional methodologies and the importance of the option-theoretic approach.

2   'Exploration must be regarded as an investment for producing future earnings from a mine constructed to exploit the orebody. The decision to undertake exploration requires the determination of the probability-adjusted NPV of the mine based on the discovery' (Mikesell and Whitney, 1987).

3   The tendency to project rising real oil prices is a post-1979 phenomenon. Prior to that oil prices were typically forecast on the assumption they would remain unchanged in nominal (or real) terms (Pollio, 1986; Streifel, 1995).

4   Of the three stages, development involves the largest investment expenditure, and so it is at this stage that the option value is greatest.

5   The option feature is the right to undo the trade, selling the underlying asset at the strike price if the market price turns out to be lower than that at expiration; the right to sell the underlying asset at the strike price is equivalent to having a put with the same strike price (Figlewski, 1990).

6   Dixit and Pindyck (1994) observe that 'since oil price uncertainty is not fully diversifiable, the greater is the perceived volatility of oil prices, the larger would be the discount rate and the smaller the value of the undeveloped reserve. However, this would understate the value of the reserve, and probably by a large amount. The reason is that it ignores the flexibility that the owner has over when to actually develop the reserve, that is, the reserve's option value. Also note that because of this option value, the greater is the volatility of oil prices, the larger is the value of the reserve.'

7   The two surveys, although conducted by the same authors, cover differ lease sales. The Siegel *et al.* (1987) results apply to a federal lease sale covering six tracts auctioned in September 1980 in which the sample company bid on each tract. The Paddock *et al.* (1988) analysis covers a lease sale held in November 1980 in which 62 tracts were awarded; of these, the authors were able to gather consistent data for only 21 of the successful bids. More positively, the sales occurred at roughly the same time, eliminating the possibility of any wide differences in price assumptions. This appears to be the case for oil prices but not for natural gas. As a first cut Paddock *et al.* value gas at thermal equivalence, but go on to argue that a straight thermal parity calculation may be inappropriate. Accordingly, they present an alternative valuation based upon gas priced at $3/mcf, with they assert corresponds to assessments used by bankers actively involved in the sale of natural gas reserves.

8   A recent study by Howell and Jagle (1997) indicates that managers are not only aware of the existence of real options but appear to value them consistently more highly than does the Black–Scholes (BS) model. One obvious explanation for this finding is that the BS model may not be applicable to the valuation of real options. The basic assumptions of the BS model are that an option contract

exists while the time to maturity is fixed. With real options, by contrast, there is no guarantee that the option will ever materialise, nor is the time to maturity certain. In other words, real options may be far more complex than their financial counterparts and thus require pricing models more sophisticated than BS to achieve an accurate valuation.

9 Less positively, Kemna (1993) notes that traditional DCF methods are incapable of providing an accurate assessment of the benefits of waiting.

10 It is worth noting that the findings of Agmon and Lessard (1977) coincide with the period in which US authorities used capital controls, including restrictions on portfolio investments in enumerated geographic areas, to help strengthen the country's balance of payments. To the extent that these restrictions prevented investors from exploiting the benefits of international diversification, favouring companies able to do for them what investors could not directly do for themselves makes sense; this is what Agmon and Lessard's results appear to indicate. Given the current high degree of capital market integration, it is doubtful whether this argument and, by extension, the empirical results that appear to support it, apply any longer.

11 Chen *et al.* (1997) report similar results: 'If two companies have the same size, bankruptcy cost, growth opportunities and profitability, but one is [a multinational company] and the other a [domestic company], our results suggest that the MNC will have a lower debt ratio than the DC.'

12 Agency costs arise when agents depart from profit maximising behaviour or principals incur costs to monitor agents so as to influence their actions (Brealey and Myers, 1991).

13 The citations for the individual studies are given in Buckley's book together with any qualifications (omitted here) that apply to the individual surveys. The tabulation is based upon questionnaire responses furnished by large US, typically Fortune 500, companies.

14 A principal shortcoming of insurance premiums, and other similar financial measures, as risk indices is they protect only the cost and not the potential profits of the investment. Accordingly, insurance premiums will understate project risk by an amount equal to the lost pay-off resulting from default (Verleger, 1994).

15 The modifier 'apparently' is inserted to reflect the fact that surveys of domestic corporate hurdle rates are not all that different from those that are applied internationally. For example, the average hurdle rate used by British manufacturing companies is roughly 17 per cent (CBI, 1994); the figures for US firms are similar. According to Summers (1987), US firms were using hurdle rates ranging from 8–30 per cent, with a median and mean of 15 and 17 per cent, respectively. Finally, Dixit and Pindyck (1995) observe that 'managers regularly and consciously set hurdle rates that are often three or four times their weighted average cost of capital'.

16 One possible reason for ignoring diversification benefits in offshore investments is that historic low correlations between the benchmark and target portfolios may have increased over time. Kim and Singal (1997), however, present evidence that highlights the relative stability of cross-country correlations: 'during the 10–year period 1982–91 the correlation coefficients between returns on the Composite Emerging Markets index and US stocks was 0.24, and increased to 0.36 for the period 1987–96. But during the two successive five year periods 1987–91 and 1992–96, the correlation coefficient actually fell, from 0.40 to 0.25.' See also Bekaert and Harvey (1995) who similarly affirm the lack of any increase in correlations *vis-à-vis* developed and emerging stock markets.

4 Project Finance

Numerous studies claim to have 'documented' the existence historically of project finance loans. If correct, this would make limited recourse structures among the oldest forms of debt known. However, as discussed in Chapter 1, a clear distinction has to be made between limited recourse loans, where repayment depends uniquely upon the cash flow generated by a single, self-liquidating investment, from other ventures comprised of a single project only. Limited recourse loans are a well-defined form of borrowing; any transaction that fails to include all of the elements unique to this structure cannot qualify as project finance. On this basis, it is arguable whether the historic record produces a single, unambiguous example of a limited recourse loan in the sense used here.

On the other hand, there is clear evidence linking project finance with the development of energy resources. According to Creath (1983) limited recourse loans were 'invented' in the late 1920s and early 1930s to provide US wildcatters with longer-term production finance.[1] During the 1930s 'as drilling became deeper and the resulting costs higher, more extended financing terms were needed... During the late 1930s and early 1940s, improved engineering techniques were developed. These improved the ability to forecast the future recovery of oil reserves, and some banks, using their own technical staff, applied these new techniques and were able to justify oil production loans with terms in excess of three years.' Commercial banks, in other words, were prepared to accept the risk that the reserves were adequate, in the sense of being able to support a level of production enabling the borrower to repay the loan solely out of the project's cash flow; by definition, the creditworthiness of the borrower was irrelevant.[2]

The scope of project finance has expanded well beyond its geographic and industrial origins. While the petroleum sector and natural resource industries remain leading users of project finance they no longer dominate the market. Lending structures, too, have evolved away from traditional export oriented projects, with their assured hard currency stream and hence the ability to escrow sales proceeds in offshore accounts. A significant part of the project finance market today is accounted for by power and other private infrastructure projects, which lack the critical features common to more traditional structures. Indeed, the concept even applies to projects without a hard currency stream in countries currently unable to borrow on their own guarantee.[3] And finally, capital constraints are no longer the driving force behind choosing project finance over more traditional debt structures; risk management objectives are now of paramount importance.

DEFINING PROJECT FINANCE

The term project finance is potentially ambiguous since it encompasses two non-mutually exclusive concepts. More to the point, it can refer to any set of financial structures used to fund project development, including limited recourse loans, the subset of financial options that are commonly referred to as project finance loans.[4] While there is fairly broad agreement as to what is meant by the term project finance, the following three quotations demonstrate that the nuances are far more significant than are the definitions themselves.

According to de Nahlik (1992) 'project finance is a way of developing a large project through a risk-management and risk-sharing approach while limiting the downside impact on the balance sheets of the developers or sponsors.' Harries (1990) asserts that 'project finance is lending to a project in which the lender expects to be repaid only from the cash flow generated by the particular self-liquidating project. The sole collateral for the loan are the assets and the revenues of such a project, except for very limited recourse to, or support by, the equity owners or other parties interested in the project.' Finally, Buckley (1996b) defines project finance as a 'highly leveraged financing facility established for a specific undertaking, the creditworthiness and economic justification of which are based upon that undertaking's expected cash flows and asset collateral.'[5]

The first of the definitions stresses motivational factors: project finance is an element in overall corporate risk management. Risks are shared with lending banks and other parties having an interest in the project, contractors or potential purchasers of project output, for example, while sponsors are able to insulate their corporate existence from any problems that may arise in connection with the performance of a single project. The other two definitions are more descriptive, focusing on those characteristics that differentiate project finance from other project financing options. In the latter two quotes, risk management is implied but not stated explicitly. The second quote favours a commercial lender perspective, while Buckley stresses the developer's perspective.

Taken together the three quotes encompass much of what has been said of project finance. In the first instance, the project is incorporated as a distinct entity, typically a joint venture company. Project sponsors are keen to maximise financial leverage – debt usually accounts for 70–80 per cent of total project capital, although some loans have been structured with over 90 per cent debt.[6] Whether this is done to exploit organisational, tax or other potential benefits connected with this type of structure is less clear. Moreover, guarantees are supplied by project sponsors to protect lending banks, but are seldom comprehensive and typically do not apply post-completion, that is, once the project has become operational. Projects routinely enjoy other levels of support provided by suppliers, purchasers of project output and government authorities, either individually or collectively. Project debt is totally separate from the sponsor's other financial obligations; repayment of the financing is limited to the cash flows generated by the project itself, while lender security is confined solely to project assets.

The interconnections that exist within project financing are summarised in Figure 4.1. At the centre is the single purpose project company. Sponsors provide equity directly to the project company in return for which they will, if the project is successful, receive dividends. An operating agreement is entered into between the sponsors and the project company, while the sponsors are obligated to provide direct support to the lenders in the early stages of the project via completion undertakings. The project company enters into separate and legally binding agreements with the lending banks, which provide the bulk of the funds used to construct the facility; with a contractor that has responsibility for building the project to the specifications detailed in the feasibility report; and with potential purchasers of

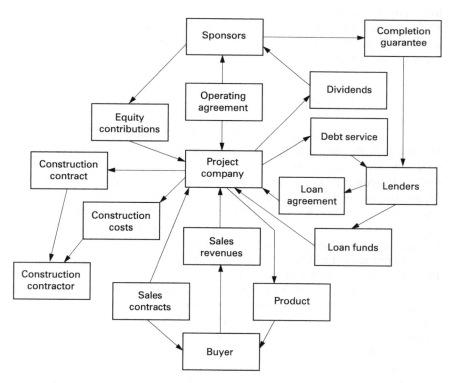

Figure 4.1 Schematic representation of project financing

project output that under most circumstances will commit to an offtake agreement only, and in exceptional circumstances to a guaranteed minimum price.

The web of interconnecting obligations and commitments illustrated here defines, as we have seen, this particular type of financial structure; less obvious is what specific interest or set of interests motivates participants to prefer this to other available options.

The traditional approach to this issue has been to compartmentalise and analyse separately the interests of leading project participants, to see whether it is possible to extract some general principles capable of providing a definitive answer to the question. To this end, research has tended to concentrate exclusively on project sponsors, commercial lenders and host governments and rightly so – the interests of all other parties, it should be apparent, are derivative however crucial their involvement may be to the project's ultimate success. Perhaps the single most significant point worth noting is that project finance has nothing at all to do with capital constraints. Major international energy, mining or power projects, as they long have been, are developed by some of the world's largest companies. There can be no doubt of the sponsor's financial capacity, thus underscoring the centrality of risk management objectives in opting to fund such ventures using project finance structures.

PROJECT SPONSORS

The relevant literature stresses precisely those themes we have already identified as the defining characteristics of project finance, namely, that sponsors are attracted to limited recourse structures because they can reallocate project risks and are able to maximise financial leverage. Several recent studies, by contrast, address such issues within the context of formal financial models thus eliminating the circularity inherent in more traditional explanations. In particular, these newer approaches show how project finance structures can be used to address agency problems, bankruptcy costs, taxation and political risk (Shah and Thakor, 1987; John and John, 1991; Chemmanur and John, 1992).

Agency problems would not exist at all if it were feasible for contractual agreements to cover all potential contingencies likely to arise in connection with the operation or management of a project. Since this is impossible, financial efficiency is enhanced by making sponsors or operators residual claimants, that is, requiring each to take a direct equity stake in the project thus linking their profits to the overall financial performance of the project. The same sort of reasoning underpins the importance of bankruptcy considerations. Consider the situation in which the costs of bankruptcy are higher for the parent company than for the project company. Integrating the project into the parent company would expose the latter to greater risk than would be the case if the project were financially segregated. Project finance effectively insulates the sponsoring company from the costs of project failure.

Given differences in tax regimes, sponsors can maximise tax shield benefits by locating the project in a higher tax jurisdiction, again arguing in favour of isolating project assets along project finance lines. Finally, non-recourse structures address political risk issues by shifting the burden from project sponsors to lending institutions or international agencies who collectively are assumed to have substantial clout and thus better able to bring greater pressure to bear on host governments should the threat of hostile action arise.

These conclusions still leave much to be desired. Isolating and analysing individual risks may seem arbitrary in that many projects normally face most or all of those just described. Indeed, it could be argued such explanations actually raise more questions than they answer. If, for example, agency concerns are of paramount importance why would single sponsors favour project finance over other loan structures? Similarly, the risk of bankruptcy obviously increases with leverage, yet it is doubtful that project finance tends to be favoured by more highly leveraged than less highly leveraged firms. Even if it that were the case, we are no better informed as to why sponsors enter into the associated complex web of contractual arrangements that are an equally integral part of project finance structures.

The same logic applies in respect of the importance of tax considerations: it would make little or no sense for sponsors to favour project finance when as sometimes happens the parent and project companies are located in the same tax jurisdiction. And finally it is arguable whether commercial banks, with or without direct supra-national participation, are able to exert greater influence over host governments than, say, consortia of leading energy, mining or manufacturing companies.

These observations are not meant to denigrate the value of project finance models. Indeed, a recent empirical study provides evidence that confirms the significance of many, if not most, of the insights contained in the theoretical literature. In their organisational choice regressions, Kleimeier and Megginson (1996) show that project size, whether measured absolutely or scaled by firm-specific variables (assets, sales or equity), is a critical factor influencing whether sponsors finance investments using project or more traditional balance sheet loans. Political risk factors appear to be an equally significant determinant of the choice of loan structure: the greater is political exposure as measured by *Euromoney's* political risk index, the more likely sponsors are to favour limited recourse loans. And finally their results suggest that multiple sponsorship does influence selection of project finance over more traditional loan structures. On the other hand, these findings appear to undercut the conceptual foundations of most project finance models, namely, that it is possible to isolate a single overarching motive for selecting limited recourse over alternative sources of debt finance.

Once we allow for the fact that project risks are multi-dimensional, there appears to be no unique reason for preferring project finance over traditional loans. Each sponsor will weigh the risks inherent in a given project, with the organisational choice being dictated by the cost and availability of alternative mechanisms capable of mitigating or eliminating those of greatest importance. For example, if fluctuations in output prices were the sole or main focus of sponsor concern, it would far more cost effective to use commodity derivatives than to structure the financing on a limited recourse basis.

Still, there is the strong presumption that project finance must be value enhancing, otherwise it would be difficult if not impossible to explain why sponsors select this over lower cost debt structures.[7] Traditional finance theory provides little reassurance on this point. In a CAPM world investors are not compensated for bearing unique risk, so the only way project finance could enhance firm value would be through a reduction in systematic risk. How project finance would achieve this outcome is by no means obvious. Similarly, if in perfect financial markets the value of the firm is independent of its capital structure, then it follows that the use of project finance can in no way be value enhancing.

On the other hand, Brealey *et al.* (1996) argue that, even in a world of perfect capital markets, project finance can 'change the state of the world in which the debt is in default. In so doing, project finance changes the expected costs of default.' More to the point, project finance can shield the firm from project default or, alternatively, protect the project from the risk of other assets defaulting; less positively, projects funded using off balance sheet structures can fail owing to the loss of the co-insurance of existing assets. Whether the implied change in expected pay-offs adds value is an open question; certainly, Brealey *et al.* make no claims to that effect.

In contrast, Buckley (1996b) maintains that project finance is under certain restrictive assumptions concerning firm valuation likely to create value. Applying the approach advocated by Lessard and Shapiro (1983), which makes firm valuation the sum of individual project cash flows less a factor that seeks to capture the impact on expected after-tax cash flows of the total risk of the firm, Buckley argues that any reduction in the risk penalty automatically increases the value of the firm. 'If we assume that the present value of a given project's future cash flows is independent of the use of project finance structures, and assume that financial markets reward risk aversion, the resulting reduction in the firm's risk penalty would increase the market value of the firm.'

The quotation actually begs the question since it presupposes that it is more financially efficient for the firm to diversify risk than leaving it to shareholders to do so, surely a contrariant assumption. It is equally plausible to argue just the opposite, that project finance averts a potential increase in the total risk penalty that would result from recourse to more traditional forms of debt. Either outcome is questionable in the final analysis, since within a total risk framework it makes far better sense to structure the project portfolio in such a way as to achieve a lower relative level of risk than to rely on financial vehicles to achieve the same result.

None of the approaches we have looked at so far provide either individually or collectively a totally convincing explanation as to why sponsors prefer project finance to more traditional debt structures. The safest conclusion we may draw from the available literature is that project finance, like other risk management structures, is intended to address the specific risk objectives of the firm. This implies that, when faced with similar objective conditions, different firms will reach different conclusions regarding the desirability or appropriateness of using project finance. Value is thus enhanced indirectly, through the consistent application of risk management principles that are fully in keeping with the preferences of the firm's shareholders.

COMMERCIAL BANKS

Commercial banks have long been providers of limited recourse loans, although it is equally true the project finance market is dominated by a subset of the largest multinational banks that historically have accounted for the bulk of such lending. Moreover, the market is partitioned into banks that are predominately arrangers and those that are mainly providers of funds. More to the point, only four of the top tier arranging banks were among the leading ten funds providers. This distinction also tends to have a marked geographic dimension, with American and European banks tending to dominate the arranging side of the market, while Japanese banks, although among the largest in the world, focus primarily on the provision of funds (see Table 4.1).

Chase Manhattan Bank is far and away the largest arranger of project loans, having put together 92 transactions aggregating $8.8 billion in 1996. Nor should this be too surprising since Chase incorporates two other multinational banks, Chemical and Manufacturers Hanover, that were major project finance banks in their own right. Additionally, Chase ranks first or second among lead arrangers in all regional markets other than the Middle East and South Asia. Among other American banks, only Citicorp and Bank of America rank in the top ten; ABN Amro, Barclays, National Westminster and Société Générale represent the leading European project finance banks.

Table 4.1   Project finance banks (millions of US dollars)

A. Ten largest global project finance loan arrangers, 1996				
Bank	*Amount*	*No. of deals*	*Average size*	*1995 position*
Chase	8 752	92	95.1	1
Citicorp	5 935	72	82.4	4
ABN Amro	4 629	64	72.3	2
Barclays	4 073	56	72.7	17
Bank of America	3 614	57	63.4	22
HSBC	2 838	39	72.8	8
Toronto-Dominion	2 819	26	108.4	7
NatWest	2 282	33	85.7	20
CIBC	2 210	30	73.7	19
Société Générale	1 863	37	50.4	3

B. Ten largest project finance providers, 1996			
Bank	*Amount*	*No. of deals*	*1995 position*
HSBC	1 853	56	10
BoT-Mitsubishi	1 817	80	2
ABN Amro	1 726	112	1
Sumitomo	1 626	120	16
Sanwa	1 573	90	11
Crédit Lyonnais	1 531	102	6
Société Générale	1 436	91	2
Chase	1 419	117	23
WestLB	1 333	68	20
Sakura Bank	1 320	100	10

Source: Sayer (1997).

Among providers, three of the top five institutions in 1996 were Japanese banks; only one American commercial bank, Chase, ranks among the ten main providers. The niche occupied by Japanese banks in the project finance market appears more closely connected with the attainment of national objectives than with the pursuit of purely commercial goals. In other words, the bulk of funds provided by Japanese banks are used to support projects beneficial to the national economy, in contrast to other multinational banks whose lending practices are driven by narrower strategic concerns.

One of the principal attractions to commercial banks of project finance over traditional balance sheet loans is that project resources are devoted to and must be used solely within the project; legally, they can not be deployed elsewhere within the firm. Once lenders have satisfied themselves as to the financial and commercial merits of the project, this feature provides an additional degree of security in respect of the project being able to meet its repayment obligations.

But project segregation has additional advantages for commercial lenders. In the first instance information costs are reduced compared with more traditional loans. In project financing lenders have only to evaluate and audit project assets rather than having to assess both project and sponsor assets as would be the case in secured financings. Equally important, bankruptcy risks tend to be low in the sense that project assets are largely tangible, so that a change in ownership is unlikely to affect project performance to any great extent (Brealey *et al.*, 1996).

This latter point can not, however, be pushed too far since the collateral value of project assets will obviously vary with prevailing market conditions for project output. For example, lenders that financed relatively high-cost tin projects in the early 1980s against the expectation the international tin agreement would remain in force over the life of the loan saw the collateral value of the loan decline *pari passu* with the large decline in tin prices that followed the collapse of the ITC in 1985. The limited recourse feature of such financings meant that sponsors were free to walk away from the project leaving it to lenders either to continue operations or else dispose of project assets to recover the original indebtedness. It is more remarkable that in several instances lenders actually sold project assets to the original mine owners albeit a price well below pre-collapse values.

Security interests can also be affected adversely by prevailing legal rules that preclude lenders from automatically selling or transferring project assets to the highest bidder; this is especially true for North Sea petroleum financings where the UK Minister for Energy has the final say in who obtains the ultimate right to the concession (Pollio, 1992). This rule was instituted to ensure efficient operation of producing concessions in the British sector of the North Sea, without regard to the impact on lender interests since the latter would obviously prefer the concession was awarded to the bidder willing to pay the highest price for the right to operate it. This is not to say banks lack an interest in operational efficiency. To the contrary, since loan repayment depends more or less totally upon efficient project management, there is an obvious interest in balancing technical with financial efficiency. But the point is that how that balance is struck is not entirely within the control of the lending banks, to the detriment of the security interest inherent in the project.[8]

It should be noted in this connection that the strategic value of the project to the sponsor means that a material adverse change in market conditions need not result in automatic abandonment, although such decisions will obviously be conditioned by, *inter alia*, potential offsetting changes in other key project parameters. The clearest evidence supporting this view is that none of the major international petroleum projects initiated in the early 1980s, whether in the North Sea or elsewhere, was bankrupted by the 1986 collapse in oil prices. This conclusion is all the more remarkable once we recall that many of these projects were conceived in the mid or late 1970s, when oil prices in real terms were far higher than they are now.

Central to this outcome was a combination of endogenous and exogenous factors that collectively helped to underwrite project cash flow. Among the former, we might note that sponsors were able to effect substantial cost economies, partly in consequence of doing away with the gold-plating that was a more or less common feature of most major North Sea projects, partly to sharply lower drilling and related expenses and partly owing to technical advances that similarly reduced development and operating costs. Among the exogenous factors, interest rates declined sharply over the 1980s, as did petroleum taxation, both providing an additional source of financial cushion to existing projects (Pollio, 1986; Alzard, 1996). In short, only under certain restrictive conditions can it accurately be said that projects have relatively low bankruptcy costs; a more accurate view requires taking proper account of the potential risks to project collateral arising from changes in the operating environment.[9]

The third, and perhaps most obvious, source of commercial bank interest in project finance is that such loans normally entail higher spreads and generate substantial fees, especially if lenders are involved in the structuring as well as funding phase of the financing, thus boosting bank earnings compared with more traditional forms of lending. Still, there is the critical issue of whether banks are appropriately compensated for the additional risks they bear within project finance structures. If banks do in fact bear equity-type risks, spreads would have to contain a quasi-equity return to provide adequate compensation.[10] Yet given the modest premiums project loans command over more traditional loans, it is doubtful whether that is the case. If so, it would provide an additional rationale for sponsor preference for project finance loans, which in effect give them access to equity but without having to pay equity premiums.

This contention, although frequently repeated, is wide of the mark since in fact banks do not bear equity risks but rather operational risks in project finance loans, for which prevailing spreads would appear to provide adequate compensation. Recall in this connection that banks assume little or no risk in the construction and start-up phases of a project – when the project is absorbing cash without generating any revenues – the true focus of equity risk. Project loans become non-recourse only *after* the project is operational and lenders are satisfied it will perform up to the standards stipulated in the project report.[11] The nature of project risks is fundamentally different in the pre-completion stage to that in the operational phase, when banks face their greatest exposure and to which loan spreads actually relate.

Recent research provides interesting insights into the characteristics of project loans, the magnitude of loan spreads and the factors that influence such margins. These data fully confirm prior expectations concerning the main features of project compared with more traditional loans (Kleimeier and Megginson, 1996). Based upon a sample of 123 project finance and 207 balance sheet loans arranged over the past 25 years, the mean project loan spread averaged 101 basis points, 32 basis points higher than spreads applying on traditional loans. The smallest observed project spread amounted to 6.25 basis points, while the highest was 300 basis points; the comparable margins for balance sheet loans were 0 and 250 basis points, respectively. Furthermore, the median project finance loan was three times larger than the median traditional loan, while the former contain a significantly larger debt component – median total debt in project loans was four times higher – and a correspondingly lower level of equity. Here, however, the margin of difference is much narrower, $70 million as against $103 million. Finally, project finance loans are exposed to far greater political risk while tenors were roughly three years longer than for traditional loans.

Kleimeier and Megginson's empirical analysis also provides interesting insights into the relative importance of sponsor, project and loan characteristics on the respective project and non-project finance loan spreads. Their analysis was carried out in two stages: in the first they regress spreads solely as function of sponsor and loan characteristics and in the second project-specific variables are introduced. Excluding project characteristics, the results are rather disappointing. The overall explanatory power of their regressions is low, with adjusted R^2s ranging from 0.16 to 0.18 and only two of the explanatory variables – guarantees and a binary variable that distinguishes project from non-project finance loans – are statistically significant at the conventional 95 or 99 per cent level. Interestingly, the project finance and sponsor guarantee variables are of roughly the same order of magnitude but of opposite sign, meaning that the two effects cancel each other in terms of impact on loan spreads. The authors do, however, concede the project finance variable may capture other influences, above all, project characteristics.

This suspicion is confirmed in the augmented loan pricing model, which shows that spreads increase with loan tenors, country risk, and dollar denomination, and decrease with the presence of guarantees and exchange risk, that is, where project cash flows are in a different currency than the loan. These results appear more robust, while the explanatory power of these regressions is roughly twice as high as in their basic model. The only finding requiring additional comment is the inverse relationship between currency risk and loan spreads, an apparent paradox which the authors side step. Perhaps the simplest explanation is that where a loan is exposed to currency risk, banks will normally require the borrower to hedge it as a pre-condition for obtaining the loan. Thus the variable may serve as a hedging proxy so that by eliminating currency risk from the financing, sponsors ought to be able to clawback the cost through a lower spread.

Lastly and most significant for present purposes is that project loans command an implicit premium of roughly 43 basis points over balance sheet loans, virtually identical to the magnitude indicated in the non-augmented regressions and more or less in keeping with the raw sample premium. This implies, of course, that banks require a premium of roughly 50 basis points, all things constant, to cover what we have described before as operational risk.

HOST GOVERNMENTS

Host governments are attracted to project finance for the same reasons as are foreign direct investors, namely, it maximises leverage, transfers risk from sponsors to lenders, is self-contained and off-budget. In simplest terms, governments can acquire needed infrastructure (roads, ports or power stations) or other local projects without having to bear any, or only minimal, cost backed up by a higher degree of technical or operational efficiency than would otherwise be the case. Where governments or local parastatals take a direct stake in the project, the equity cash call is normally limited to roughly 12.75 per cent of project capital cost, equal to the product of the government's majority interest (51 per cent) and the typical equity percentage (25 per cent).

The fact that many governments choose or are obliged by local law to hold a majority interest in the project raises several contentious issues, not the least being whether such involvement is compatible with the efficient management of the project. In one sense this issue is irrelevant since without majority ownership the project would never have been implemented; in other words, it is the price foreign sponsors have to pay to realise the investment. It is important, however, to differentiate ownership from management since in the more usual case it is the foreign investors that have day-to-day responsibility for operating the facility, regardless of the prevailing ownership structure. However, it is not unusual for state authorities to exercise close and direct operational control in those instances where qualified and experienced local personnel exist.

Indeed, such considerations often times dictate the ultimate shape of the project as the contrast between build-own-transfer and build-own-operate structures attests. In the first instance, the foreign investors own and manage the project until the initial investment is recovered; thereafter the project reverts to the state without any continuing obligation to the project on the part of the original foreign sponsors. In the second structure, ownership and control are vested in the local authority from the outset, the degree of foreign operational control varying from project to project and dependent upon local competencies. Infrastructural projects raise the interesting possibility that state participation, far from being unwelcome, may address risks that otherwise could frustrate efficient realisation of the project. If the state owns the project outright or knows that it will be transferred to local control in due course, the incentives for providing continuing project-related support may be far stronger than would otherwise be the case.

Host governments are also able to create value from resources that previously remained unused or wasted, again without having to incur any or only modest expenditure. A classic instance of value creation is the use of natural gas that formerly was either re-injected or flared. In the early 1960s, oil exporting countries typically flared more than half their annual associated gas production; the comparable figure for non-OPEC countries was less than 5 per cent (Adewole, 1992). By 1990, OPEC countries were able to reduce significantly the volume of flared gas, to around 10 per cent, through export, petrochemical conversion or power development, with an attendant favourable impact on global pollution. More recent data would, no doubt, show that the flared percentage has continued to decline. Still, the difference *vis-à-vis* OECD gas producers, while having narrowed appreciably over the past 30 years, is still high, underscoring the potential for further gas-based projects.

But the benefits to host countries go further than that. Based upon local resource potential, governments can build project portfolios that address both industrial and risk-return issues. For example, natural gas producers may include an aluminium smelter in their project slate, knowing that the low correlation between aluminium and, say, natural gas export cash flows means they can achieve a significant reduction in the volatility of state (royalty, tax and dividend) revenues.

There are additional benefits in terms of local industrial, economic and financial development, not the least more effective project management. Foreign sponsors will in most instances introduce state of the art technology, organisational or administrative structures and heightened awareness for safety and environmental concerns. For many projects, especially those located in the higher income emerging markets, equity or debt tranches may be reserved for local investors, with the knock-on effect of promoting the development of local financial markets.

Reflecting an awareness of the potential industrial and financial benefits connected with greater foreign direct investment, host governments virtually everywhere have revised or modified prior restrictions on foreign involvement in the local economy, modified or harmonised tax regimes in keeping with the new investment rules, and have embraced privatisation and other ownership structures that limit public expenditure, build-own-transfer or other similar financial vehicles, for example (Rose, 1992; Walde, 1992).

The same question discussed above, concerning whether project finance creates value, can be raised here as well. After all, governments have options other than, or supplemental to, project finance to develop local projects, each entailing different financial options and management control that may cater more directly to local requirements. Project finance along with privatisation represents one end of the spectrum. In both cases, finance for and management of the project are the responsibility of the private sector. At the opposite extreme are nationalised industries, where the state provides the capital for and is the sole owner of the project. There are, of course, intermediate options, service contracts or leases, for example; in the case of service contracts, the state provides finance while the private sector is responsible for project management, while for leasing just the opposite division of responsibility applies.

The case for project finance normally rests on the assumption that private investors are better able to quantify project (and, indirectly, social) benefit, provide more efficient operational control and manage the attendant risks than is the state, either directly or through local parastatals. In other words, capital is valued in terms of its opportunity cost, ensuring that scarce financial resources are allocated optimally to the benefit of economic efficiency. However, governments can normally provide finance for local project development on far more favourable terms than can private investors, which typically require returns substantially higher than the official cost of capital. Thus, as several commentators point out, the choice of structure ought to balance the gains connected with private management with lower project funding costs. If this argument is accepted, it is doubtful whether project finance enjoys a clear advantage over alternative organisational forms which favour direct state involvement.

Financial analysts are unanimous in condemning these conclusions, arguing that such thinking represents a thinly disguised defence of public subsidies or else fails to comprehend the true cost of official capital. Alexander (1995), assessing the criteria that should be used by governments providing funds to state enterprises, concludes that 'important in this decision-making process is the determination of the cutoff point above which returns are deemed economic and below which any funds should be counted as aid. For a private company this cutoff point is its cost of capital... There is no reason to judge public companies differently.'

Brealey *et al.* (1996) go one step further, questioning the premise that governments do in fact enjoy a lower cost of capital than does the private sector. '[Project risks] do not disappear when the project is financed by the government. If cash flows from the project are unexpectedly low and do not suffice to service the debt raised by the government to finance the project the shortfall is met by taxpayers, who play a role similar to that of equity-holders in a private sector company (but without the benefit of limited liability). Indeed, because it seems likely that the capital markets share risk better than does the tax system, the cost of capital for the government could well be *higher* than for corporations.'

PROJECT FINANCE AS A STRATEGIC OPTION

The preceding discussion highlights the limitations inherent in traditional approaches to assessing the value-enhancing potential of project finance structures. Generally speaking, the best such approaches can achieve is to define the conditions under which limited recourse financing either makes sense or does not. Value creation potential is thus derived obliquely: if it makes sense to use project finance structures then, implicitly, the value to the sponsors must be greater than if alternative debt options had been pursued. We can, however, avoid the resulting circularity by posing the issue somewhat differently, in particular, by focusing on the option-like characteristics inherent in project finance. In effect, the methodology being advocated here is an extension of recent research that argues that project structure is as – or more – important than the project itself (Pindyck, 1991; Dixit and Pindyck, 1994).[12] In contrast to more general formulations of the problem, which seek to identify the existence of, or the potential to create, value-enhancing real options within existing project structures, our main objective is to show how financial structures, above all, limited recourse loans, can be used to the same effect.

A central feature of project finance is that upon completion project risks are transferred from project sponsors to the lending banks. Should objective conditions that determine project values change, sponsors have the option of either continuing to repay the loan or else default. The example given above, of tin operations that ceased being economic once prices fell following the collapse of the ITC, well illustrates the principle. If the project had been carried on the sponsors' books, they would have been obliged to absorb the entire loss connected with project failure; the use of project finance effectively shielded the investors from having to incur such penalties. In fact, sponsors were free to repurchase their assets, at the lower implied market valuation, once abandonment had occurred.

In the projects-as-options literature, the principal emphasis would have been on the additional flexibility, and hence value, created by the sponsors having an implicit right of abandonment. The pay-off to the abandonment option is max(salvage value – project value, 0), the value of which can be determined in keeping with traditional options valuation models. In other words, the right to abandon is seen as being equivalent to the sponsors having a put option to sell the project for its salvage value, even if the present value of prospective cash flows was less than that amount. It is worth noting that having exercised the option, the firm would still have to bear the financial consequences of abandonment since it is implicitly assumed the project had been financed solely with equity.

But if instead the project was financed with limited recourse debt, the pay-off to abandonment now becomes max(project value – the value of outstanding debt, 0), that is, the option will be worth whichever is greater, the project value minus the value of the outstanding loan or zero. Since the value of project assets is identical to the market value of project debt and equity, it should be clear that sponsors will only exercise the option if project equity is positive. In effect, sponsors have succeeded in transferring the risk of project failure to the lending banks. By electing to use project finance, sponsors in essence acquire an option post-completion *not* to reclaim the assets from the banks having a legal charge over them.[13]

More formally, the use of limited recourse debt is equivalent to the sponsors selling the assets of the project to the lending banks in return for (1) the proceeds of the loan, (2) a call option to repurchase project assets from the lenders with an exercise price equal to the face value of the loan,[14] and (3) a lease enabling the sponsors to use project assets over the life of the loan.[15] The option can be valued in keeping with the Black–Scholes valuation model, an exercise that has the additional advantage of isolating the importance of those features that are more or less unique to project finance loans. The Black–Scholes call option solution for project equity, E, is:

$$E = VN\{\ln(V/X) + (r + \sigma^2/2)T/\sigma\sqrt{T}\} - e^{-rT}XN\{\ln(V/X) + (r - \sigma^2/2)T/\sigma\sqrt{T}\}.$$

In general form, the value of project equity can be expressed as:

$$E = E(V, X, T, \sigma^2, r).$$

Since by definition the value of project debt, D, is equal to the value of project assets less project equity, the value of project debt can be determined according to:

$$D = VN\{-\ln(V/X) - (r + \sigma^2/2)T/\sigma\sqrt{T}\} + e^{-rT}XN\{\ln(V/X) + (r - \sigma^2/2)T/\sigma\sqrt{T}\},$$

or, more generally, as:

$$D = D(V, X, T, \sigma^2, r).$$

Table 4.2 defines the variables used in the model and presents the partial derivatives of E and D, respectively, with respect to the factors shown in the first column of the table. With the exception of V, the partial effects of the listed factors on the option values of E and D are, as expected, of opposite sign: changes in the variables that influence option prices redistribute project values; they do not enhance the value of underlying project assets. By contrast, an increase in the value of project assets, V, directly increases the value of *both* project equity and debt; owing to the attendant increase in

Table 4.2   Impact of a change in model variables on project equity and debt

Variable	Description	Change in the Value of Project	
		Equity	*Debt*
V	Value of project assets	+	+
X	Face value of project loan	−	+
T	Loan tenor	+	−
σ^2	Variance of annual rate of return on project assets	+	−
r	Risk-free rate of interest	+	−

debt coverage, there is a corresponding reduction in the probability of default and accordingly an increase in the market value of project debt. Similarly, an increase in promised debt repayment – equivalent to an increase in the lending banks' claims on project assets – raises the value of project debt. Since sponsors are residual claimaints, there must be a corresponding reduction in the amount of cash flow available for distribution to the equity holders and hence a decline in the value of project equity.

On the other hand, an increase in the loan tenor or the riskless rate of interest reduces the present value of the debt, simultaneously lowering and raising the market value of project debt and equity, respectively. The same conclusion applies in respect of an increase in the variance of returns on total project assets, which can be thought of as a proxy for the variance of after-tax cash flows available for interest payments and for distribution to equity holders. Since the lending banks have a maximum payment, X, which they can receive, 'an increase in the dispersion of possible outcomes increases the probability that the value of the [project's] assets will be below the promised repayment, thereby increasing the probability of default, thereby lowering the value of debt [and] increasing the value of equity' (Smith, 1986).

One of the more crucial features of the options valuation model for present purposes is, as noted, that the value of the call on the company's assets is positively related both to the volatility of underlying project assets and time to maturity. Now, one of the crucial determinants of the equity cash flow stream in a levered project is the amount of debt used to finance the assets, such that the value of the option will increase with increases in financial leverage. The value of the option is also an increasing function of the time to maturity, which equates to the tenor of the loan. The longer is the maturity of the debt, the more valuable is the call option for repaying the loan and hence for retaining control of project assets. Available data confirm the main predictions implied by the Black–Scholes model with respect to project finance loan characteristics: projects financed with limited recourse debt tend to be more highly levered and have longer tenors. According to Kleimeier and Megginson (1996) the mean (median) tenor of project finance loans is 11.8 years (10 years) compared with 8.1 (8.0) years for traditional balance sheet loans.

The options approach also provides additional insights as to the value-creating potential of limited recourse finance. Leverage, as we have seen, does not effect the market value of project assets, only the relative value of underlying project debt and equity. More to the point, higher debt levels, by increasing volatility, are consistent with higher equity valuations; in effect sponsors have succeeded in creating value by moving from low risk to high risk investments, through a process known as risk shifting.

Commercial lenders, to be sure, are aware of the options-like characteristics embedded in project finance loans. They obviously, therefore, have a strong interest in seeing to it that sponsors exercise their call option to repay the debt, and so will be keen to set the exercise price as low as possible relative to the market value of the project's assets. This may seem contradictory, since by definition project financings are highly leveraged transactions. On the other hand, the complex web of contractual commitments and restrictive covenants characteristic of project finance loans effectively restricts the ability of sponsors from in any way altering the underlying structure of the project to the detriment of lenders' interests, an outcome underwritten by the fact that project assets are dedicated solely to project use. Equally important, commercial lenders will normally set loan values using extremely conservative criteria, meaning that the actual level of debt will typically fall short of the project's true debt carrying capacity.

In the case of reserve-based loans, which are used to finance petroleum development, banks normally insist upon coverage ratios of 2 to 1 (Pollio, 1986, 1993), that is, the ratio of cash flow remaining in the project must at all times be twice the outstanding loan balance, *and* 1 to 1 coverage at the bank's low price case. It should also be noted that the price premises upon which cash flow potential is evaluated tend to be equally conservative. Banks typically set output prices at prevailing market prices and allow for rates of escalation no more rapid than the forecast rate of inflation. In effect, cash flow is assessed at prices *below* the implied real price of oil once the secondary low price constraint is taken into account.

CONCLUSION

In this chapter we have tried to show that the more usual explanations for choosing project finance over other debt options, either individually or collectively, are unconvincing. True, these approaches have succeeded in identifying a wide range of benefits connected with the use of project finance, but these are typically ancillary, not paramount. The main attraction of limited recourse loans in our opinion is the risk management features, in the narrow sense used above, that are inherent in such structures. Within an options framework, limited recourse finance represents an attempt by sponsors to exploit the financial counterpart to the real project options now widely recognised in the capital budgeting literature. This framework has the additional advantage of explaining aspects of project finance, the importance of leverage, longer tenors or the complex set of contractual commitments and loan covenants, for example, that formerly were understood, incorrectly as it turns out, in other ways.

Notes

1   The earliest recorded loan secured by a mortgage on an oil reserve was made in 1928. Tenors on early production loans averaged three years (Creath, 1983).
2   According to Creath the willingness to provide petroleum finance was based upon the lenders' ability to quantify the project's reserves and hence production profile. The omission of any reference at all to price risk is significant. This can be explained in the first instance as resulting from the application of officially sanctioned production controls in Texas, a leading oil producing state, following the earlier huge increase in regional production capacity with its attendant depressing impact on prices. Later, the US government tacitly supported the efforts of the major integrated oil companies to 'stabilise' prices throughout the early post-World War II era. In 1973, the mantle passed to the Organisation of Petroleum Exporting Countries, which administered oil prices until they collapsed again in mid-1986. The breakdown of market control is said to have contributed to a significant increase in price volatility with its resulting negative impact on future industry growth. As we argued in Chapter 3 this interpretation is self-serving and open to serious question, as the recent scale of industry investments attests. The simple fact is that world oil prices have always been extremely volatile; it was the 1930–86 period that was exceptional (Verleger, 1994).
3   The provision of project finance in such instances abstracts from the problem of sovereign risk through the reallocation of such risks, including convertibility and regulatory risk, to official third parties, export credit agencies, for example. In addition, most of these projects are co-financed with the World Bank or IFC as well other national development agencies, such as the Commonwealth Development Fund or the Asian Development Bank. See Smith (1995) for a discussion of the Hub Power project in Pakistan where all of these elements are addressed in the financial package.
4   In view of the fact that project finance can mean different things to different people, Chapman (1995) recently suggested the term should be dropped in favour of 'limited recourse *senior* bank debt' (italics in the original). Although eminently sensible, this definition lacks the simplicity inherent in the more traditional description. We shall use the term project finance in the sense defined by Chapman.
5   Significantly, neither author mentions that project finance loans are senior debt.
6   Within the energy sector, the sole exception is refinery investments. Owing to the higher volatility of refining margins, commercial lenders will normally require a significantly higher level of equity, typically committed on a contingent

basis, to provide cash flow during periods of negative margins. Thus for the Rayong and Star refineries the overall (direct plus contingent) equity commitment was of the order of 60 per cent (Baxter, 1996).

7   To some commentators, costs of funds are deemed central to the financial viability of the project (Chapman, 1995). This conclusion is too one dimensional; other project parameters, production levels, input costs or output prices, for example, might be expected by and large to exert a far greater impact on project success. The relative importance of key project parameters is determined through the application of sensitivity analysis. Banks will normally require sponsors to address key project risks as a precondition for providing limited recourse financing. Funding risks normally fall into that category.

8   Most commentators now agree that a security interest in the project is taken mainly for defensive reasons, to prevent third parties interfering with the project or obtaining interests in project cash flows or assets that take priority over lenders' security interests (Gewirtz, 1995).

9   As shown in Chapter 5 there is strong evidence favouring the view that sponsors stick with projects even if they are currently losing money by doing so. This behaviour can be rationalised within an options-theorectic framework: as long as investors are satisfied upside potential exists it would be unwise to abandon the project. Abandonment is irreversible, foreclosing all possibility of access to project upside. On the other hand, abandonment makes sense where sponsors see little or no upside potential, as was true following the collapse of the International Tin Agreement in 1985. To maintain prices at target levels, the buffer stock manager was obliged to accumulate ever larger tin inventories. Since liquidation of these inventories at one go would have further destabilised the market, stocks might be expected to be sold off only relatively slowly. The resulting supply overhang meant it would take a very long time before the market returned to balance (Anderson and Gilbert, 1988).

10   Wood (1980), among others, takes the view that project finance fills the grey area between debt and equity.

11   Given the critical importance of completion, it should come as no surprise that this issue accounts for the lion's share of time spent in negotiations between sponsors and lending banks.

12   One of the key implications of this view is that traditional discounted cash flow methods may result in projects being wrongly rejected. Allowing for the options inherent in most project structures could lead to a complete reversal of the ranking of the financial merits of alternative, mutually exclusive projects (Kulatilaka and Marcus, 1992).

13   The reason why sponsors have a call, and not a put, option is that project assets have already been mortgaged ('put') as security for the loan.

14   More accurately, the borrower acquires a set of sequential call options equal in number to the number of contractually committed repayments.

15   The value of the lease equals the value of the collateral less the value of the debt less the value of the call (Smith, 1986).

5 Identifying and Managing Project Risks

As we have seen, project finance is concerned with the quantification and management of project risks. According to the more traditional view of project finance, risk management lies at the core of sponsor interest in favouring this over rival debt structures. Project risks can be identified precisely, carved out and allocated to third parties best able to bear them. Sponsors are thus able to shed risks that they are either unable or unwilling to bear. Without such arrangements lenders would require investors to cover any residual risks that might impair the project's ability to meet debt service obligations, in effect transforming a limited recourse loan to something more closely resembling a secured financing.

The options-theoretic view makes risk management an even more critical component of this type of structure. In contrast to more traditional explanations, within an options framework the onus for managing project risks shifts to lenders whose interests are safeguarded via various contractual agreements and restrictive covenants that are an integral part of the loan agreement. On this view, sponsors seek to maximise the value of project equity by increasing volatility. This is accomplished by financing the project with the maximum amount of debt that lenders are willing to provide, hence the crucial importance of leverage to this type of structure. In other words, risk shifting replaces risk management as the overarching objective of project sponsors. Commercial lenders accordingly are keen to manage project risks in ways that ensure that borrowers will always exercise their option to repay the indebtedness.

PROJECT RISKS: A SUMMARY

Risk within a project context means different things to different participants. In simplest terms, it might be defined as the chance of a loss. To sponsors this could mean the project failed, for any number of reasons, to generate an appropriate investment return; for commercial lenders it refers to the possibility that the loan will have to be written off. An alternative, equally acceptable, definition of risk is the divergence of actual from expected results. In practice this definition is little different from the one just given, but instead views project risk relatively rather than absolutely. In terms of outcome, the two are of course equivalent.

Another meaning is the possibility of an undesirable contingency occurring. Project economics are typically evaluated in keeping with the base case, albeit the one most likely of a set of possible outcomes. Sensitivity analysis is used to assess the impact on project economics of an undesirable contingency occurring; as part of this process, sponsors will normally seek to develop an estimate of the probability of that particular event happening. Seen this way, a project's cash flow profile is a weighted average of various possible outcomes with the weights proportional to the probability of the individual risks occurring. Implicit in this view is that specific risks are known, or at least knowable, and hence can be quantified. The final definition of risk worth considering within the context of project analysis is exposure to mischance. Unlike the previous definition this one is predicated upon an unknowable and hence unquantifiable event occurring. An earthquake in an area where no such event was previously recorded would be a good example of this phenomenon.

All of these definitions apply equally within a project context, and each affects the interests of all project participants. The resulting exposure varies over project life, however, with sponsors being at greatest risk should an unfavourable contingency arise prior to completion and lenders post-completion. The conversion from guaranteed to non- or limited recourse status, however, may create an additional category of risk not comprehended within the definitions just given and uniquely affecting the interests of commercial lenders, namely, moral hazard. Moral hazard may be defined as a situation where the element of care changes in keeping with changed objective circumstances. Since the financial consequences of project failure differ in the pre- and post-completion phases, it could be argued that a corresponding shift in the degree of care exercised by project sponsors may occur once they no longer have complete financial responsibility for project performance.

The possibility that project finance structures create moral hazards for commercial lenders cannot be dismissed out of hand but neither is it worth pushing the point too far. At the most basic level of analysis, we might expect the commercial interests of project sponsors to transcend the project's stage of development. If project economics are vindicated, there is no reason to suppose sponsors will manage operations any differently over project life, since to do so would unnecessarily destroy investment value. In any event, we may reasonably assume that the complex legal structure of the loan agreement itself provides commercial lenders with sufficient protection against this as well as the more specific set of risks that such provisions are formally designed to address.

Various techniques are used to analyse project risks. With the exception of projects that depend upon new and hence untried technology, there normally would be a sufficiently large number of existing similar projects whose historic performance provides quantitative benchmarks against which new project operations can be measured. In other words, investors and commercial lenders are able to develop robust probability estimates of key project parameters and hence assess their significance for project performance. It should be noted in this connection that lenders normally discount new technology to the extent it is essential for meeting specific production or cost targets (Clifford Chance, 1991). From a lender's perspective, new technology is equity risk and accordingly must be covered directly by project investors.[1]

Apart from statistical analysis, the basic techniques used by commercial lenders for analysing project risks can be subsumed under three main headings: credit principles, portfolio diversification and insurance. The first is the most straightforward in that commercial banks over time will have developed coverage and other credit tests that determine the maximum loan value a given project can command. Credit principles evolve continuously, as they must, to take account of changes in the circumstances upon which the guidelines were initially based. To illustrate the point, we might note that commercial lenders responded to the increased volatility of industrial metals prices in the 1970s, and the concurrent disappearance of long-term, fixed-price contracts in favour of market sensitive pricing, by adopting revised rules toward mine financings that favoured low over higher cost producers. More specifically, major mine finance banks then restricted the provision of project finance loans to producers whose cost structures were at or below the industry quartile. The principle here was that only these operations were likely to survive periods of prolonged price weakness and thus generate sufficient cash flow to repay new or outstanding loans.

This modification was based upon two assumptions, one of which subsequently proved correct, while the other did not. The correct judgment was that industrial metal prices were unlikely to recover strongly from then-depressed levels; nor as a corollary were lenders any longer prepared to accept the industry view that metals prices had to rise more or less steadily in real terms over the long run if new mine production was to be forthcoming. Project economics, in other words, were evaluated by applying unchanged real prices to a long run supply price that equated to the industry's average historic real price (Fitzgerald and Pollio, 1982, 1984).

The second and incorrect assumption was that cost structures are immutable: a given producer's position on the industry cost curve is fixed forever by technical factors. Prevailing cost pressures did cause many mines to shift the nature of their production from high to lower cost operations. It was during this period that leading US copper companies, for example, switched to electrowinning solvent extraction, and in the process transformed their cost structures from being among the highest to among the industry's lowest. It is arguable whether this transition would have occurred if banks had not modified antecedent lending rules. The financial pressures resulting from the new, more stringent credit requirements could have caused high cost producers to consider technologies that otherwise might have been resisted. On the other hand, depressed market conditions meant that most operators then were having to compete on the basis of cash operating costs rather than price. Such pressures would inevitably have provoked widespread production shifts, partly in response to the behaviour of rival producers and partly to insure continued survival.

Two important conclusions emerge from this brief survey. First of all, it reinforces the observation that commercial lenders regard established credit principles as the best safeguard against project risk. And second, it demonstrates that loan portfolio decision rules evolve in keeping not only with changing market conditions but also in response to altered perceptions concerning long-run industry dynamics. The two may, but need not, be interconnected although in all fairness in the case to hand the former did in fact stimulate a reconsideration of the latter. Commercial lenders were thus among the first to recognise the fallacy of inherent natural resource scarcity, at least as applied to metal and mining projects. It is less clear why they failed to apply the same lessons to energy projects. We noted earlier that the price premises used by commercial banks, like those applied by the industry itself, were based upon escalating real prices. The second oil shock of 1979–80, which triggered the paradigm shift, still dominates the economic evaluation of petroleum projects to a great extent. While the paradox cannot easily be explained it does raise interesting questions concerning the consistency with which lenders apply basic credit principles across industries whose long-run behaviour is ultimately driven by the same set of fundamental influences.[2]

Within specific industrial sectors, banks normally apply internal credit requirements consistently. There are, however, circumstances in which lenders are prepared to waive or modify some or most of these rules. Financings based on amended or altered credit principles are known as policy loans. Policy loans are rarely made, the exceptions are intended to promote what are deemed to be compelling commercial interests, to improve tier position with an existing client or to develop new business relationships, for example. The reluctance to make policy loans derives from the simple fact they create unnecessary risks. Within the energy sector, for example, most of the losses experienced by banks following the 1986 oil price collapse can be traced more or less uniquely to such loans; most energy financings based upon traditional credit principles, by contrast, survived the price decline.

Diversification plays the same role among commercial lenders as it does in the other areas we have analysed, namely, to minimise the risk of the overall loan portfolio and thus stabilise interest income. Commercial banks accordingly diversify their project portfolios both across industrial sectors and geographic regions, and periodically rebalance their existing exposure, like other investors, to ensure overall portfolio variance is minimised. In theory, political, like industrial, risk corresponds to unique risk and thus can be eliminated largely or entirely via diversification. In practice, bank appetite for political risk varies from country to country and accordingly strict country limits apply. For projects in emerging market economies with no or limited access to international financial markets, banks will normally require the elimination of political risk as a precondition for making the project loan. For emerging countries that can borrow on their own guarantee, the resulting exposure will be priced in keeping with prevailing market spreads, although again available lending capacity will be determined in keeping with existing country exposure limits.

Official political risk insurance, provided by export credit agencies among others, is used to eliminate the more common types of political risk, expropriation or inconvertibility, for example. Project specific risks, which normally cannot be covered by official insurers, are addressed via the private market and banks typically require that such insurance is in place prior to drawdown, with the proceeds of the policy assigned to the lending banks. As we shall see in due course, the provision of insurance varies with the type of risk being underwritten; in some instances, considerable insurance capacity exists, while in others the market is fairly thin and the premiums correspondingly high. It should, however, be clear that the willingness of lenders to bear project risks is confined, by and large, to a small subset of financial risks, generally relating to interest or exchange rate exposure, where banks are the logical source of the coverage.

MAJOR PROJECT RISKS

Since project finance is ultimately concerned with ensuring the sufficiency of cash flow to repay both principal and interest, the logical place to begin our survey is with an evaluation of the impact that specific project risks can have on individual cash flow items. Table 5.1 presents a highly stylised cash flow statement, together with a summary of the principal risks affecting the constituent entries. Project revenue is the product of production volumes and prevailing output prices. Output can be affected by both operational risks and *force majeure*, so-called acts of God that are outside direct project control. The output of most projects is sold in competitive markets where prices may be subject to considerable volatility. Commodity prices, for example, fall into this category.

For vertically integrated projects, prices may be market sensitive or, more likely, based upon internal transfer prices. Whatever benefits transfer pricing may yield in terms of revenue stability,[3] project integration could magnify project risks since the cash flow stream is now a function of the performance of a second project exposed to risks similar to those applying to the first. With prices determined exogenously, the main source of

Table 5.1   Cash flow risks

	Cash flow	Main risk
	Production	Operations, *force majeure*
(times)	Price	Market
(equals)	*Revenue*	Volume
(less)	Operating cost	Operations, transport
	Interest	Funding
	Depreciation	Completion
	Overheads	Operations
	Taxes	Political
(equals)	*Net income after taxes*	
(plus)	Depreciation	Completion
	New debt/equity	Syndication
(less)	Capital expenditure	Engineering
	Loan repayments	Funding
	Working capital	Operations
(equals)	*Change in cash*	

volatility in project revenues can be attributed to production risks. Operating risks (together with transport risks) also affect production costs, overheads and working capital requirements. Funding risks relate principally to debt service, while tax charges are a function of political risk. In short, project cash flow volatility results from numerous individual risks, some of which are pervasive while others affect individual entries only.

Broadly speaking, project risks fall into one of five principal categories: intrinsic risk, production and technical risks, economic risks, political risks and management risk. Intrinsic risk relates mainly to unforeseen problems that emerge only as the scope of the project is more precisely delineated. Typical examples of this type of risk would be ore grades that are inferior to the estimates given in the project feasibility report, or petroleum reserves that are lower than originally assessed and hence insufficient in quantity to repay the debt portion of the financing. Production risks by and large are technical in nature; in de Nahlik's (1992) words it is the risk that 'the dignitary invited to launch the project cuts the tape and nothing happens.' Included in this category would be technological risks, *force majeure*, construction delays, cost overruns, supply and quality of either production inputs or outputs and infrastructure problems. Economic risk encompasses a wide range of risks connected with the operation of the project and the market or markets for the project's output. More specifically, we might include in this category price and various financial risks, such as currency or interest rate exposure, inflation, project returns and financial projections generally.

Political risks are largely self-explanatory and cover any actions that might be taken by host governments that impair project economics. These would include such things as expropriation, inconvertibility, the imposition of new taxes or the withdrawal of previously agreed subsidies, implementation of tariffs or other barriers affecting the project's ability to source needed equipment or supplies on international markets, unilateral changes to, or abrogation of, key contract provisions, and so forth. Political risk is normally regarded as applying primarily or exclusively to projects in emerging markets, although it should be clear that such risks can and do affect projects in the OECD countries. For example, the failure of local regulatory authorities to grant adequate or timely price increases for, say, infrastructure projects could be just as devastating to project economics as are the more familiar instances of political risk. The differences, in other words, are more a matter of kind than concept. The final category encompasses the experience and reputation of all parties involved with the project, including the personnel involved in project operations, whether expatriates or locals, and continuity. Because of their obvious importance to project success, it is worth considering each of these risk categories in somewhat greater detail.

A natural resource project's ability to generate adequate cash flow to service project debts and to earn for investors their required rate of return depends crucially upon the quality and quantity of recoverable reserves. Project reserves are categorised either as proved or probable, the distinction relating mainly to the degree of certainty attaching to the quantities of project output that can be produced at prevailing or prospective market prices. The higher the market price, the greater will be the volume of reserves that qualify as proved. In some instances, the distinction is more a matter of semantics than probabilities. For example, a project may have reserves that can be produced economically at current prices but for which no sales outlet presently exists; natural gas projects are a prime example of the importance of this effect. Reserve evaluations rely upon assessments developed by independent consultants with recognised industry expertise and are audited over project life. In general, project finance markets favour hydrocarbon over mineral reserves, especially in respect of recoverability.

From a commercial lender's point of view, only those reserves that can be produced under current market conditions are taken into account in setting a project loan value. A closely related principle is that more is preferred to less: lender interest in a given project varies in direct proportion to the excess of reserves over the quantities required to service project debt. Reserve adequacy is measured by the so-called reserve coverage ratio. Project cash flow is evaluated by applying the lender's price premises to production forecasts, underpinned by the reserve estimates, with the resulting stream discounted again using the lender's preferred discount rate.

The resulting cash flow series is divided into required debt service payments to derive the coverage ratio. The conventional rule of thumb is that the coverage ratio should be at least twice the contractual debt service payments or, equivalently, that half of potential cash flow should remain in the project after loan repayment. For this reason the coverage ratio is sometimes called the half-life test. Projects having lower coverage ratios may still be bankable, but the basic loan structure will have to be modified to take account of the resulting greater risk. One way to do this would be to have sponsors guarantee a minimum level of marketable production (McKechnie, 1990; Clifford Chance, 1991). Alternatively, the loan could be divided into two tranches, with the first and shorter tranche financed on a limited recourse basis, while the second and longer maturity would require recourse to the sponsor should cash flow prove insufficient to repay the loan (Buckley, 1996b).

A closely related form of reserve lending is the borrowing base revolver. This structure provides for annual or semi-annual determination of the loan value based upon the project's cash flow potential. Borrowers are required to repay any portion of the amount outstanding under the facility in excess of the maximum loan value. Conversely, lenders will normally increase the loan value in keeping with enhanced cash flow potential resulting from higher prices or higher production volumes. It should be clear that such structures can and have exacerbated financial difficulties during periods of falling prices. The sharp drop in oil prices that occurred in 1986 triggered substantial loan repayments and a concurrent reduction in the scale of borrowing capacity, thus compounding antecedent financial pressures.

Raw material and supply risk applies equally to natural resource and manufacturing projects. For example, aluminium projects are predicated upon the availability of both alumina and electric power supplies, highlighting the importance to project economics of potential raw material shortages and input price risks. Availability risks can be addressed either through the provision of long-term supply agreements or stockpiling essential raw materials or through some combination of the two. Not surprisingly, lenders normally favour projects that source most or all of their production inputs from within the sponsor's supply chain, ideally from a dedicated facility.

Price risk is another matter. Such risks can, in principle, be addressed within supply agreements negotiated with other operations owned by one or more members of the sponsor group. The supplying plant may be prepared to offer matching long-term contracts that price inputs using transfer rather than market prices, with one plant in effect subsidising another to maximise returns on the sponsor's overall project portfolio. Failing that, lenders will evaluate project economics by applying market prices to required production inputs. If supply and input price risks are deemed excessive, lenders may require sponsor or third party guarantees to cover such risks. For some inputs, price risk can be hedged directly in forward or futures markets. In marginal situations, that is, where the willingness to lend to the project is contingent upon satisfactory resolution of such risks, lenders will normally insist that sponsors mitigate supply price risks.

In some instances, sponsors may elect to do so voluntarily, against the reasonable expectation that, by having mitigated or eliminated a key project risk, lenders will improve the terms of the financing. At minimum, sponsors will expect to be compensated for the additional costs incurred for addressing such risks if they are to consider the option. Compensation can take various forms including, *inter alia*, lower interest rate spreads, higher loan values or the right to swap equity for debt, post-completion (Pollio, 1990).

One of the key reasons lenders are willing to finance projects on a limited recourse basis is the expectation the project will conform in all essential respects to the profile given in the project feasibility report. Completion risk addresses these concerns. There are two principal dimensions to completion risk: mechanical completion and economic completion. Mechanical completion relates to the ability of the project to perform up to the technical standards detailed in the feasibility report with respect to both the quantity and quality of project output. Economic completion is concerned with the operator's ability to keep production costs to within stipulated limits and meet other well-defined financial targets. In both instances the benchmarks against which project performance are measured are those described in detail in the project feasibility report. The feasibility report is compiled by an independent expert with considerable experience evaluating similar projects and, as the name implies, provides a comprehensive assessment of the technical and economic viability of the proposed venture. Feasibility studies are expensive and can account for up to 5 per cent of total project cost (Mikesell and Whitney, 1987).

The project report is based upon four categories of data: (1) geological data; (2) location data; (3) engineering data and (4) market data. The first item relates to natural resource projects and the data generated would cover, *inter alia*, the quantity and quality of reserves and other pertinent information connected with the discovery or deposit. For pipeline or tolling projects the relevant data consist of likely plant throughput; for transportation projects, the analysis would focus on passenger or traffic flows. Location data cover things such as the costs of acquiring the project site, construction and development costs, the availability and cost of services to the project site including the provision of water, transportation and communications, political considerations, the availability of local labour and so forth. Engineering data include optimal extraction methods, equipment requirements, maintenance and replacement schedules, capital and operating cost estimates and environmental considerations. Market data include an analysis of the existence of accessible markets, whether local and/or foreign, for project output or services, prospective demand in those markets, competitive factors and output price projections over the life of the project.

From a lender's standpoint the importance of the feasibility report is that it documents all aspects of project performance and establishes tolerance levels for all relevant project variables. Warranties and representations, with respect to project performance, are backed up by stringent mechanical and economic completion tests. Inherent in the formulation and implementation of such tests is the requirement that if the project fails to meet set standards by a certain date, the sponsors will repay the loan. Mechanical completion tests apply both to plant construction and physical operation, to determine whether the project is capable of operating to design specifications. The period over which such tests apply varies and depends upon the nature of the perceived risk. Certifications for well-established functions are usually of fairly short duration; for others the tests can and do run for much longer.

Economic completion tests are equally critical and cover both cost and financial performance, including the project's ability to meet pre-set cash, working capital and cash flow generation tests. The former are typically satisfied in one of two ways. The first is through assurances by the shareholders or sponsors of the project that the project company will always have sufficient funds to complete and operate the project. Alternatively, project sponsors may agree to satisfy certain specific predetermined financial ratios. The centrality of technical and economic completion to the project's ability to generate the cash flows needed to cover debt service obligations is reflected in the fact that such certifications occupy far and away the lion's share of attention in loan negotiations and documentation.

Operating risk overlaps with, but is conceptually distinct from, both completion and raw material supply risks to which it relates; time is the decisive factor, differentiating this from the other sources of project risk. More to the point, operating risk is concerned mainly with the types of difficulties projects may encounter over the life of the financing and includes such problems as escalating labour, energy or equipment costs, transportation bottlenecks, operating problems that derive from poor engineering or design work, high maintenance costs owing to corrosion or wear, adverse exchange rate movements and managerial incompetence.

Among the risks just noted, perhaps the most important relates to managerial ability. Lenders attach as much – or greater – weight to the project's operator as they do to purely technical issues, on the principle that even the best designed project will not produce the desired financial results if it is poorly organised and managed. Additional complications arise in connection with joint ventures, where differences in managerial ability and experience are often masked by the operational structure. These problems are particularly acute for joint ventures involving parastatals, where local law may require the replacement of expatriate managers by appropriately qualified nationals. What matters in the final analysis is not the formal organisational structure of the joint venture company but the *de facto* division of responsibility among the industrial partners, in respect of both operational and financial matters. Technical excellence unaccompanied by financial discipline is a recipe for disaster.

The second area of major concern to commercial lenders is the behaviour of operating costs. Since cash flow is by definition net revenue, the failure to keep project costs under firm control can have a negative impact on project economics as big as revenue shortfalls. In some instances, costs are subject to direct control; in others, they depend upon exogenous factors, which may or may not be susceptible to hedging. In this latter connection we might note that mining and metal processing operations, whose economics are sensitive to energy costs, were naturally severely affected by the second oil shock of the late 1970s, to the detriment of project debt service obligations. Smith and Walter (1991) cite the example of a lateritic nickel mine in Queensland, Australia, where the rise in energy prices increased fuel costs from 10 to 50 per cent of the project's operating costs resulting in four debt reschedulings, a massive overhang of unamortised debt and deferred interest. Not surprisingly, most banks that participated in the project lending syndicate were obliged to write-down their loans.

Smith and Walter go on to note that another significant source of operating risk is the quality and availability of local labour supplies, especially for projects located in developing countries. It is implicit in their discussion that local labour pools tend to be inefficient in comparison with those in the OECD, raising operating costs to the detriment of project economics. The best way to address this risk, they argue, is through training, astute labour relations and sourcing expatriate workers.

Limited available data, however, appear to belie Smith and Walter's contention. Poten and Partners (1991), for example, present wage costs and labour productivity for selected LNG sites referenced to the US Gulf Coast. True, the figures show that labour productivity on average is lower in emerging market economies than in the OECD, differences that are not always compensated by lower wage costs. Still, if these figures are to be believed, labour productivity in some of the industrial countries included in the sample is not much different from that of emerging economies, with the impact on costs magnified by substantially higher local wage rates. Wage costs in Norway, for example, are more than twice those applying in the US Gulf Coast, yet Norwegian labour is about 25 per cent less efficient. The result is that real wage rates in Norway are roughly three times higher than in the United States. The same general conclusion applies to Australia. On the other hand, Russian wage rates are significantly lower and labour productivity higher than in either Norway or Australia. Real wage rates in Russia are accordingly only two-fifths the US level and an even smaller fraction of Australian or Norwegian labour costs.

Market risk refers to the possibility that the project may not be able to sell its output at a price that collectively covers operating costs, debt service charges and the investors' required rate of return. There are in fact two dimensions to market risk. The first is that the project will not be able to produce the quantity of output or to the standard of quality stipulated in the project feasibility report; this latter consideration means that even if quantitative production targets are being met purchasers of project output have the right to refuse delivery or, alternatively, that the price received by the project will be discounted compared with the prevailing market price. This risk, known as operating risk, was discussed previously.

The second and ultimately far more serious risk is that sponsors will have completely misjudged the medium- and longer-term demand for project output, with a consequent weaker price profile than initially projected. Cash flow will, accordingly, be lower than anticipated to the detriment of both the sponsors' ability to earn their required rate of return, and the project's ability to service its debt obligations. Within a limited recourse framework, both of these risks may compromise the interests of the lending banks. In the first instance, sponsors would be well within their legal rights to abandon the project, leaving it to the banks either to continue operations or to dispose of project assets to repay the outstanding indebtedness. The importance of the second consideration depends upon whether the problem surfaces pre- or post-completion. If at the former stage, then any cash flow deficiencies will have to be made good by the project sponsors, in keeping with the prevailing guarantee structure; if in the latter phase of the project, the outcome for the lending banks would be no different from abandonment, with identical financial consequences.

We have already argued that this risk is particularly acute for natural resource projects, above all, oil and natural gas investments. This is partly because of the changes in the structure of the petroleum market that have occurred over the past decade. However, a more fundamental reason is the consistent misapplication of natural resource pricing models that bias oil price forecasts upwards and hence imply a far more favourable cash flow profile than would result from using prediction methodologies that do not assume resource scarcity. As long as development costs are below prevailing market prices, resource development will be profitable (Adelman, 1993); the assumption that escalating real prices are a necessary condition for undertaking petroleum investments is redundant.

The final three major risk categories are political, financial and *force majeure* risk. Political risk has already been discussed, where we concluded that for projects in emerging market economies, lenders will normally require traditional risks to be mitigated via the provision of official or private insurance. For projects in the industrial countries, where political risk typically equates to regulatory risk, lenders will normally accept such risks on the not unreasonable assumption that countries having an interest in promoting private investments will, by failing to provide adequate or timely rate increases, effectively undermine investor interest in providing the desired investments.

Financial risk is another matter. Projects whose costs and revenues are fixed in terms of currencies other than the one used in the financing are obviously exposed to foreign exchange risk. Moreover, since most projects are priced in terms of a spread over the London Interbank Offer Rate (Libor), funding costs too can vary conceivably quite considerably over the life of the financing. In most instances, borrowers can hedge most or all of the resulting exposure by borrowing in the same currency as that generated by project sales. Residual foreign exchange risks can be managed either through recourse to short- or long-dated forward exchange contracts or derivative instruments such as swaps, options and so forth.

Similar instruments exist to manage interest rate exposure including forward rate agreements, options to hedge short-term risks and swaps, caps and collars for longer-term exposure management. As a general matter, lenders will require sponsors to eliminate such risks through the use of any or all of the instruments just enumerated. Where similar instruments exist, commodity price (output) and supply (input) risks can be managed in exactly the same way; where, by contrast, such instruments do not exist, or cannot be matched to the tenor of the loan, the resulting ongoing exposure may require direct sponsor support as a precondition for obtaining project financing.

Finally, *force majeure* risks, commonly referred to as 'acts of God', are outside the control of either project sponsors or lenders; typical examples of acts of God are floods, tidal waves, hurricanes, typhoons, and so forth. The occurrence of such an event, quite obviously, impairs performance in either the construction or operational stages, increasing financial charges (by increasing capitalised interest) in the first instance and in the second to the detriment of the project being able to service its financial obligations. The resulting suspension of operations can be of short or long duration, so that the ultimate impact on project economics will vary directly with the length of time project operations are affected. In most cases, lenders are unwilling to assume *force majeure* risks, meaning that the resulting exposure will have to be covered either by project sponsors or the private insurance market. Private insurance that addresses *force majeure* does exist, but coverage tends to be limited in amount and the associated premiums accordingly are quite high. How such exposure is addressed may be the responsibility of the project sponsors or, alternatively, may be divided between sponsors and the lending banks.

North Sea oil and gas projects provide an interesting case study of the way responsibility for managing project risks, unknowable and hence unquantifable for the earliest developments, was initially divided between sponsors and lenders and how, with growing experience, the risk trade-off changed. North Sea development antedated the first oil shock of 1973. However, the combined impact of the five-fold increase in prices and the seizure of producing concessions in the leading oil exporting countries gave a significant fillip to the development of oil resources in secure provinces, above all those in the OECD region.

With production now established for almost three decades it may be hard to grasp how difficult and complicated early project developments were. Most obviously, there is the harshness of the North Sea environment, meaning that in many instances sponsors were obliged to develop and perfect new technologies to exploit the region's resource potential. Under most circumstances, as we have seen, new technology is anathema to commercial lenders, who regard such exposure as equity risk and thus solely for the account of project sponsors. Moreover, being a new region, there was no production history; accordingly there was initially no basis against which reserve or production estimates could be measured.

Lenders faced numerous additional problems. Commercial exploitation of the region's hydrocarbon potential depended upon successful development of a new and costly infrastructure, a problem compounded by the lack of a local construction industry and, by extension, the absence of an experienced indigenous labour force. In fact, in the early stages of development, the bulk of the workforce was recruited in the United States. Among the particular logistical issues that had to be faced, oil had to be landed by tanker, loading from an offshore buoy near a production platform or by subsea pipeline and run through to an onshore terminal. Nor was production quality uniform but rather varied from field to field, creating the risk that it would be extremely difficult to define a reference price against which regional production could be valued. In the event, once production commenced, spot and forward markets developed rapidly and one particular blend – Brent – eventually emerged not only as the benchmark regional crude but, following the 1986 oil price collapse, the marker against which most internationally traded crudes are now valued.[4]

Finally, there were (and are) two legal requirements that initially greatly complicated the smooth application of the project finance structures, widely used to develop petroleum resources in the United States, to the North Sea. First of all, project sponsors in the United States typically own the reserves that secure project financings. In the North Sea, by contrast, the Crown is the sole owner of mineral resources, meaning that there is no possibility of lenders taking over project assets in the event of default. To be sure, concessions can be sold or transferred by one operator to another but only with the prior approval of the Minister of State for Energy, making lender interests subordinate to those of the state, precisely the reverse of the situation applying in the United States. In the case of natural gas, the same problems naturally applied but with the additional complication of their being only a single buyer, British Gas, and without an agreed sales contract there was no possibility of field development.

Against this backdrop, it might reasonably be asked why commercial lenders were prepared to fund North Sea development using project finance structures. The short answer is also the obvious one: in the early stages of regional development banks relied upon guarantees provided by parent companies to local affiliates on whose behalf the financings were being arranged; some also acquired a direct equity stake in the producing fields (de Nahlik, 1992). The financing of BP's Forties Field, among the first regional developments, well illustrates the terms and conditions that governed the earliest North Sea projects (Brealey and Myers, 1991). The Forties Field was developed with a $945 million loan provided by a syndicate of 66 international banks to Norex, a wholly owned BP affiliate company. Norex in turn paid over the sum to BP Development, another BP subsidiary company charged with responsibility for constructing the project. Norex's payment to BP Development took the form of an advance payment. BP Development agreed to repay the advance in oil, against a back-to-back purchase commitment from BP Trading, yet another affiliate company, with the sale proceeds used to service the original bank loan. Payments to Norex were spread out over ten years, although the rate of repayment depended *inter alia* on actual production rates.

It should be clear from the preceding discussion that banks assumed reserve risk only; all other project risks were covered, directly or indirectly, by BP. Two other specific risks addressed in the financing were (1) the possibility that the field might be depleted at a faster rate than expected, to the detriment of the project's longer-term cash flow generating potential; and (2) *force majeure*. The first contingency was covered by an undertaking from BP to make good any debt repayment deficiencies (that is, the difference between project revenues and debt service obligations) by agreeing to pay the difference into a 'reclaim account'. *Force majeure* risk was defined as a situation in which no oil was produced after six years, even though the underlying reserves upon which the financing was based were adequate to service project debt. In which case, Norex could claim repayment from a restitution account guaranteed by BP.[5]

A closely related question is whether the apparent success that banks had in funding North Sea oil development can be attributed to their ability to analyse and manage project risk successfully or whether it had more to do with luck than anything else. Castle (1986) was among the first to question the rationality of the high level of interest shown by both investors and lenders in major North Sea projects. His analysis, covering 23 regional project developments, shows that, for the sample as a whole, capital cost overruns were nearly double the original estimates; that cumulative production through the early 1980s was 40 per cent lower than predicted; and finally that operating costs were 142 per cent higher than initially estimated and were likely to be 625 per cent higher over field life. The only saving grace was that project cash flow was 40 per cent higher than projected, thanks to the two oil shocks of the early and late 1970s, which resulted in significantly higher prices than originally anticipated. According to Castle's analysis, project success had little or nothing to do with the ability of banks to quantify and manage project risks; luck was the decisive factor.

Using the same data, Pollio (1986) showed that the evolution of lending to North Sea projects reflected a far clearer understanding of project risks than implied by Castle's analysis. The hypothesis tested was that lender attitudes followed a well-defined learning curve. Modifications to antecedent lending rules occurred against the backdrop of a steady improvement in operating performance, the result of growing experience and confidence in managing large-scale regional developments. The test consisted of relating capital cost overruns, cumulative field production and operating costs, first, to the date projects were initiated and second to project size and complexity as proxied by investment cost. The evidence shows that later North Sea projects had lower capital and operating cost overruns and smaller production errors than did earlier developments. And second, while complex projects faced more serious problems than did less complex developments, the resulting correlations were generally low and of doubtful significance. But perhaps the most compelling evidence casting doubt on Castle's conclusions is that the 1986 oil price collapse, and the persistence of relatively low prices since then, has yet to undermine the economics of any of the projects included in his sample.

We have already explained in Chapter 4 why this occurred; here it is sufficient to reiterate the point that the production history of the North Sea vindicates lender rationality in respect of providing finance for large-scale projects in an unknown and unproven oil province. This experience goes far towards explaining the current high level of interest in providing finance for oil and gas development in the former Soviet Union and elsewhere. As in the early stages of North Sea development, the legal and fiscal regime evolved *pari passu* with industry growth. In the North Sea, taxation rules changed often, alerting lenders to the risks connected with volatile fiscal regimes, issues lenders are again having to face, hopefully more knowledgeably, in the former Soviet Union.[6]

MITIGATING PROJECT RISKS

Lender exposure to project risks varies over the life of the financing, but is at its peak just after operations commence. During the engineering and construction phase, the first stage of project development, lenders disburse funds in keeping with the schedule detailed in the loan agreement; these drawdowns are, however, fully secured or covered by sponsor or other third party guarantees that effectively shield the lending banks from the consequences of project failure. The second – and from a lender's standpoint – riskier stage is the start-up phase, where the project has been completed, is producing and thus is generating cash flow, a portion or, conceivably, the entire amount of which is now dedicated to repaying the lending banks. It is during this phase that lenders will have satisfied themselves as to the technical and economic merits of the project by ensuring that all the completion undertakings stipulated in the loan agreement have met or exceeded the standards contained in the project feasibility report. Once these tests have been satisfied, the loan converts from being a secured to a limited or non-recourse financing one.[7]

In the final, operational, phase financial responsibility lies primarily or completely with the lending banks, with the degree of exposure dependent upon whether the loan was structured on a limited or, less commonly, non-recourse basis. On the other hand, it is during this last major phase that the project is generating the cash to repay the loan, so that the balance outstanding is also declining steadily. Once the loan is repaid, tantamount to project sponsors having reacquired title to project assets, the lending banks have no residual obligations towards either the project or project sponsors.

Lenders are normally unwilling to allow borrowers to draw down the loan facility until after project equity has been committed; in some instances, loan disbursements are made *pari passu* with equity infusions. The principle applying here is that lenders are always keen to ensure that sponsors have something at risk before loan proceeds can be applied towards project construction or development. Superficially, the principle appears prudent and sound, but in fact is a relic of past lending practices; the real safeguards lie, of course, in the legal and other contractual undertakings contained in the loan agreement, which effectively underpin the financing. The fact that the scale of project equity is a factor of relative rather than absolute importance is proven by the fact the equity requirement set by banks varies from project to project. One of the key determinants of the equity requirement is, as we have seen, a determination by the lending banks that the investment is 'rich' enough in virtually any project state to repay the initial indebtedness.

Still, there is the strong presumption that commercial pressures may result in banks setting project equity requirements at lower than optimal levels. There can be no denying that competitive factors do influence the terms and conditions governing the financing, although it is arguable whether such pressures manifest themselves as a partial or complete waiver of established credit criteria. True, this does happen as we noted before, but only rarely and even then solely in keeping with what are perceived to be compelling commercial interests. In the more usual case, competitive pressures affect loan parameters *ex post*: once the project is up and running and to the extent new loans for similar projects in the same sector are being negotiated on more favourable terms than the pioneer financing, borrowers will normally be able to refinance their existing loans on terms corresponding or, quite possibly, superior to those currently applying in the market.

One of the key roles performed by limited recourse lending is to mitigate project risks by reallocating them to those best able to bear them. Different theoretical perspectives suggest different motives for doing so, but whichever paradigm is ultimately applied, risk management remains of critical significance. It is important, therefore, to define which third parties are best able to cover leading project risks. To that end, Table 5.2 divides and subdivides project risks into the principal analytical categories discussed above. The table also indicates the specific hedging tool best suited to addressing individual project risks, and, finally, considers which participant (or group of participants) is the logical source of the coverage.

Most of the entries in the table are self-explanatory. In the construction phase, the bulk of project support, as might be expected, is provided by the sponsors or the contractor designated to build the project. For major projects, contractor guarantees are normally satisfactory; where there are questions as to the contractor's creditworthiness, an extremely unlikely occurrence in respect of major international projects, lenders will require that sponsors provide the ultimate guarantees. Other than sponsors or contractors, the only other third party likely to be involved in the early stages of project development are insurance agents, which provide coverage against interruptions resulting from *force majeure* events.

Of the remaining three risk categories, two – political and financial risk – are pervasive. While sponsors remain the single most important source of risk coverage in the operational phase, suppliers of raw material inputs or other project services (transportation, for example) or purchasers of project output provide an important degree of support that helps to ensure both smooth project performance and, by extension, the availability of cash flow in amounts sufficient to service outstanding loans. Again, *force majeure* surfaces as an issue in the operational as in earlier project stages, with the provision of adequate coverage a *sine qua non* for commercial bank funding of project development.

Financial risks, by contrast, are covered entirely by third parties, either financial institutions or suppliers that provide the instruments to hedge the negative impact such risks might have on project economics. The potential adverse impact of inflation on project performance is addressed within the framework of long-term contracts entered into between the project and its suppliers or purchasers. Finally, provision against certain key aspects of political risk again falls mainly on project sponsors; more traditional country or political risks – expropriation, nationalisation, contract abrogation, and so forth – will normally be covered by national or supranational agencies or the private insurance market.

Table 5.2 Project risks, hedging tools and sources of coverage

Risk	Hedging tool	Source
Construction and completion risks		
Supply and availability of raw materials and building materials	Supply or pay contract	Supplier
Adequate communication	Project's network	Sponsors
Contractor's performance	Feasibility study	Sponsors
Force majeure	Insurance	Insurance agency
Cost overruns	Completion guarantee	Contractor
	Standby credit	Lenders
Delays	Completion guarantee	Contractor
Operational risks		
Energy supply	Long-term supply contract	Energy supplier
Output	Take and pay contracts	Purchaser of output
Transportation	Long-term transportation contract	Sponsors
	Project's transportation infrastructure	Sponsors
Operator performance	Feasibility study	Sponsors
	Compensation arrangements	
New technology	Licensing agreement	Licenser/sponsor
Conflicts of interest among sponsors	Inter-sponsor contracts	Sponsors
Resources	Feasibility study	Sponsors
Force majeure	Insurance	Insurance agency
Financial risks		
Exchange rates	Options, futures, swaps, and so on	Financial institutions
Inflation	Long-term supply and output contracts	Suppliers and purchasers
Interest rate	Fixed-rate loan, interest ceilings, interest rate derivatives	Financial institutions, lenders
Political risks		
Availability of licences and permits	Good working relationship with government	Sponsors
Expropriation	Participation of local sponsors, international agencies, lenders	Sponsors
Country risk	Feasibility study	Sponsors
	Insurance	Insurance agency
Sovereign risk	Feasibility study	Sponsors

ALTERNATIVE RISK MANAGEMENT STRATEGIES

In the present section we are concerned with the alternative ways key project risks can be mitigated or eliminated. In particular, we contrast contract structures, the traditional means by which project risks have been (and still are) addressed with financial or commodity derivatives, instruments that can be used to the same effect. Since risk management entails cost, there must be some reason or set of reasons why project sponsors are prepared to incur such expenses. The most obvious is that without having addressed specific risks of direct concern to lending banks, the latter either would be unwilling to provide debt finance for project development or, alternatively, require a far higher equity commitment from the sponsors. More positively, active risk management can lead to an improvement in the terms of the financing. Finally, the hedging programme can be executed independently of the financing or as an integral part of the loan agreement. Where a hedging programme must be in place as a precondition for providing finance, the issue is decided by the lenders.

The following stylised analysis is intended to illustrate how derivative instruments can be used to address key project risks and why, accordingly, commercial lenders are prepared to improve the terms of the financing once such structures are in place. Here we are concerned with the risk profile of an independent power project, say, a co-generation facility, located in the United States; as we shall see the principles are perfectly general and apply *mutatis mutandis* to similar projects regardless of their location. Co-generation projects involve the simultaneous production and sale of electric power and steam, possibly, although not necessarily, to the same utility company. For example, steam could be sold to local authorities for heating purposes or to industrial users, chemical plants, among others, that use it as part of their production process. For present purposes the distinction as to ultimate purchaser is of no importance; what matters are the terms and conditions under which project outputs are sold.

By way of background, independent power production in the United States did not develop to address a specific need but rather in response to federal incentives, above all, the Public Utility Regulatory Policy Act. Co-generation, however, is a cost-effective way of generating electric power, so that utility companies, financially hard pressed at the time the legislation was enacted in the late 1970s, were not entirely unenthusiastic about allowing new entrants to cover a portion of prospective power requirements rather than having to incur the investment expense themselves.

One of the critical features of the US legislation is that utility companies are required to purchase surplus power either at a fixed rate or at the utility's avoided cost. Avoided cost involves complex calculations that in effect take account of the financial savings that accrue to utilities from having unaffiliated generators supply power that otherwise would have had to be provided out of the utility company's own capital budget. For our purposes, the key feature of the utility market is that all or a significant portion of project output is covered under a long-term, fixed-price agreement, up to 15 years in some cases; in many instances it is also possible to dispose of the steam on the same or similar terms.[8]

This feature stands in marked contrast to most other commodity projects, where output is virtually always sold at market-sensitive prices; offtake, by contrast, may be covered under long-term supply agreements, although it is arguable whether 'long-term' now means very much more than the coming 12 or 18 months. On the other hand, independent power producers may be subject to input price and supply risk. Accordingly, abrupt cost increases or the failure of a producer or group of producers to supply the fuel input, typically natural gas, are as inimical to project economics as are output price or offtake risks.

From the perspective of project finance markets, independent power projects present a risk profile similar in some respects to that of natural resource investments, but with some significant enhancements that go a long way towards explaining the high level of interest shown by lenders in these types of projects. The most obvious commonality is technical risk; there is also the possibility that construction costs will exceed the levels established in the project report, while both are subject to operational risk, that is, once the facility is up and running it may fail to meet standards set with respect to production or cost levels or both.

The key differences, as we have seen, relate mainly to the arrangements covering offtake: power is typically sold under long-term contracts to purchasers having a legal obligation to buy the output at predetermined prices, in direct contrast to the situation faced by sponsors of more traditional commodity projects. Since tariffs are set by regulatory authorities – the Federal Energy Regulatory Commission in the United States or the Office of Electricity Regulation in the United Kingdom – there is always the risk that the principles upon which output prices are currently set will be revised, even radically, in keeping with a change in the prevailing political or industrial climate. Indeed, there is the strong possibility that purchasers will actively seek to modify or terminate existing purchase agreements so as to avoid the excessive financial burden that existing contractual commitments can impose upon purchasers in the face of changed market conditions.

An important example of this phenomenon, all the more significant given the project's location, is to be found in Pakistan. The success of the Pakistani government's private power initiatives appears to have resulted in the construction of too much generating capacity. Accordingly, local authorities are looking for ways to escape current high cost purchase agreements – the present tariff structure was established in the first instance to attract private developers – by attempting to exploit legal ways out in existing sales and purchase contracts. The approach currently being pursued is both encouraging and dangerous. It is encouraging in the sense that, on past practice, we might have expected local authorities to have repudiated unilaterally current high cost contracts or, alternatively, to have enacted legislation revising the existing tariff structure, either option constituting a classic instance of political risk.

By operating within the existing legal and contractual framework, Pakistani authorities appear determined to avoid sending the wrong signal to foreign direct investors while simultaneously trying to revise power rates paid to foreign developers that now appear well above market clearing levels. Whether the two objectives are compatible is another matter.[9] The risks to Asian power development, of course, run deeper than that. Owing to the economic difficulties being experienced by many leading east Asian economies, the main focus of recent international private power and infrastructural development, these risks are bound to become more acute to the detriment of established power projects. Moreover, we might reasonably expect recent economic and financial pressures to have a depressing impact on the scale of future regional power and other private finance initiatives.[10]

Apart from regulatory risk, which applies equally to power investments in both developed and emerging market economies, there is the narrower issue of the creditworthiness of purchasers of project outputs and input suppliers, another feature common to both independent power and more conventional natural resource projects. As we shall see one of the main attractions of using commodity derivatives to hedge project risks is that they more or less do away with the creditworthiness issue. To see why, consider traditional hedging strategies, that is, selling project output under long-term sales contracts that stipulate both offtake and price, or entering into long-term supply agreements that mitigate availability and price risks. Indeed, the two agreements may be linked. The fuel supply agreement may be a precondition for obtaining long-term sales and purchase agreements, on the principle that the project, by always having access to fuel supplies of the appropriate quantity and quality, should be able to meet its sales commitments.

The creditworthiness issue is of course relevant to both sets of transactions. Any deterioration in the credit standing of the purchasers of project output will undermine the contracts upon which project cash flows and by extension the financing are based. With respect to the fuel supply agreement, there are two critical issues – price and availability effects. The relative importance of each depends principally upon the pricing formula contained in the supply agreements. To the extent that the project is able to pass through automatically higher purchased fuel costs, then availability is likely to become the main focus of supply risk. The qualification is intended to highlight the importance of regulatory risk noted above. If, on the other hand, the project is unable to pass along higher fuel costs, price risks will compound availability risks to the detriment of project economics.

Leaving supply availability risks aside, a not unreasonable assumption in view of the large number of domestic producers of both natural gas and fuel oil, the main focus of project risk is likely to centre on the cost of fuel inputs. In the first instance, assume the generator is unable to pass along higher input costs. Project cash flows will be adversely affected more or less in direct proportion to the increase in fuel costs relative to base case assumptions. Under these circumstances, lenders are likely to view the project as being exposed to a high degree of risk and may therefore require sponsors to address the issue via one of the two hedging options previously outlined. Failing that, lenders will lower the project loan value or, equivalently, require a higher level of equity, an outcome opposite to what sponsors seeking project finance would have wanted.

Nor should there be any presumption that banks are indifferent between the two options: creditworthiness issues suggest the superiority of the use of financial derivatives over a long-term fuel purchase agreement. Note, too, that even if the sales and purchase agreement has a fuel cost pass through clause, project economics would be enhanced if operators were able to eliminate the potential negative impact of higher purchase gas costs on cash flow. The fact of being able to use hedging instruments to generate a higher level of cash flow than would be the case if the project were exposed to input price fluctuations, means that projects with hedged fuel supply costs should be able to command a higher loan value than an unhedged operation.

There are various ways the project could hedge its fuel purchase costs. From a lender's standpoint, the ideal situation would be for the term of the hedging contract to correspond identically to the loan tenor. In this way, lenders are safeguarded over the life of the financing against the risk of a price spike occurring that would significantly lower cash flow and hence reduce the loan coverage ratio to less acceptable levels. One way to address this risk would be through two back-to-back contracts, the first, a traditional fuel purchase agreement with prices set at market clearing levels, and the second a cash-settled hedging transaction (a swap) that fixes fuel supply costs over the life of the loan at a level that ensures cash flow sufficiency in keeping with target coverage ratios.

Under these circumstances, the project always pays the market price for its fuel inputs. If the market price is above the fixed swap price, the counterparty to the swap (the swap dealer) pays the project the difference between the two prices; if the opposite price relationship applies, then the project pays the difference. True, the project is unable to benefit from any favourable fuel price movements, a cost that is counterbalanced by the certainty of having its fuel costs capped at a level set by the fixed swap price and thus having the ability to achieve the sponsor's desired degree of leverage. There are of course more flexible options that can be tailored to meet the specific requirements of project sponsors, while satisfying the risk management objectives of the lending banks; whatever the form, the outcome would be little different from the one just described.

In the example just given, the project was able to hedge its input costs thus eliminating a key project risk. For many natural resource projects it is possible to hedge both output prices and input costs, thus enhancing project economics in comparison with the uncovered alternative. An obvious example is an aluminium project, where electricity charges, the single largest component of variable cost in producing primary aluminium, can be fixed as can output prices. Refining projects provide a second example of the benefits to sponsors of being able to fix both input (crude) costs and output (refined product) prices. Since processing margins are extremely volatile, lenders are unwilling to fund refining projects on terms normally applying on limited recourse loans, while borrowers typically avoid project finance since the equity requirement – direct and amounts that must be committed on a contingency basis – is far too high in relation to the sponsor's target equity level. In the two examples just given, hedging results in a far more favourable and certain cash flow profile, so that lenders can provide a more attractive financing package than would be the case if leading project risks remained unhedged.

THE CONTRACTUAL FRAMEWORK

The legal structure of a project financing, other than the loan agreement itself, can be organised into one of five basic categories. The first and most obvious relates to the contractual agreements that govern relations among the project sponsors. The most important of these is the shareholder agreement which, among other things, defines the reciprocal rights and obligations of sponsors with respect to business policy, financial exposure and construction and operation of the project. Concession agreements, the second broad category of legal arrangements of concern here, cover the relationship between project sponsors and governmental authorities. The importance of these agreements derives from the fact that all sponsors, regardless of where the investment is located, require licences or other permits from local authorities to construct and operate a project. These authorisations include such things as building licences, mining concessions, water rights, local procurement requirements, the waiver of tariffs or other trade restrictions on the importation of project equipment, and so forth. For projects located in emerging market economies, sponsors and the host government will normally enter into basic agreements that cover all of these and related issues.

Because of their centrality in influencing a sponsor's decision to proceed with a given investment, concession agreements warrant further consideration. From the sponsors's perspective, the importance of the concession agreement lies in the fact it creates a legal obligation with respect to both construction standards and completion dates. Under the terms of the concession agreement, the project company is generally required to issue warranties covering the construction work undertaken and to pay liquidated damages in the event the project is not completed in keeping with the originally set schedule. Sponsors normally pay a concession fee for the right to build a project, while the project company enters into legally binding commitments with the host government that it will operate and maintain the project over a defined period of time and to predetermined standards. There are, finally, other agreements that confer upon local authorities the right to terminate the concession in certain circumstances, for example, as a result of the project company having breached the concession agreement in some way, and that cover the disposition of project assets upon expiration of the agreement (Clifford Chance, 1991).

The third broad contractual category relates to construction contracts or development management agreements that the project company may enter into. Master construction contracts apply only in those circumstances where the project lacks the resources or expertise to build the project. Where the project company is organised as a finance vehicle, the usual arrangement is to enter into a comprehensive turnkey contract with a single contractor or construction consortium; the contractors also have responsibility to sub-contract work, either to local or international construction companies. An alternative arrangement is for sponsors to appoint a project management company vested with responsibility for arranging, with a number of contractors, a series of construction contracts with the project company to build the facility. In some instances the choice of sub-contractors is dictated by the terms of the concession agreement itself, especially where local operations exist and are capable of meeting international standards with respect to equipment supplied or quality of construction work. The establishment of international standards of quality means that the contractor has (in principle) an automatic right to source the needed supplies or equipment on international markets, normally without having to incur prevailing customs duties

The final category of contractual arrangements we are concerned with covers sales and purchase agreements; these agreements are central to projects being able to access limited recourse debt in the amount and on terms generally sought by the sponsors. Sales and purchase agreements exist in those situations where project output is not disposed of in either spot or retail markets. In the latter cases, cash flow potential is evaluated at expected market prices; fixed-term contracts, now more or less uniquely based upon market sensitive pricing, may however contain a premium to prevailing market prices to reflect the economic value to the purchaser of the long-term commitment.

Project sponsors themselves are, in many instances, the sole or major purchasers of project output and thus provide what could be regarded as an unconditional cash flow guarantee. The terms under which project output is purchased, even when covered by agreements with project sponsors, are variable and can take several forms, including throughput agreements, take-or-pay contracts, with or without unconditional purchase committemnts (so-called hell-or-high-water clauses), take-and-pay contracts, take-if-tendered contracts, and so forth. Quite obviously, the stronger the purchase obligations are, and the higher the prices to be received by the project, the stronger is underlying cash flow support with an attendant favourable impact on the potential scale of the lending commitment.

For obvious reasons commercial lenders prefer take-or-pay to the alternative purchase contracts. Take-or-pay contracts require the buyer to purchase a minimum quantity of project output whether it is currently needed or not.[11] The importance of this type of contract lies in the fact that if the purchaser fails to take the contractually committed minimum (assuming, of course, the project was able and willing to deliver), then the purchaser is obliged to pay as if the agreed quantity had been acquired; of course, if a take-or-pay payment is made, the purchaser is usually entitled to acquire an equivalent volume in the future and at no cost. In virtually all cases, take-or-pay contracts stipulate offtake only; prices are normally market sensitive, that is, set in keeping with formulae agreed between the project company and the purchaser of project output, that take explicit account of prevailing market conditions. For example, the agreed price of liquefied natural gas (LNG) in the purchaser's home market might include the cost of competing fuels, say, heavy fuel oil or coal. Alternatively, it might relate uniquely to the price of other LNG imports, confined either to contract purchases or, less restrictively, to a weighted average of contract and spot imports. In some rare instances, purchasers will agree to a minimum floor price, set at a level that (ideally) ensures the project is able to cover both the investor's required rate of return and the project's debt service obligations. Minimum purchase or floor prices can still be found in recent LNG projects, Qatargas, for example.[12]

Generally speaking, long-term, fixed-price contracts were central to early LNG investments where the risk of project failure – magnified even then by the high price of liquefied gas compared with both refined products and pipeline supplies – was effectively underwritten by purchasers of project output. Purchasers were prepared to concede minimum prices because they were confident the higher prices could be passed through to final consumers. Nor was this confidence misplaced: as long as local or regional gas markets remained subject to continued tight regulatory control, there was no reason to suppose that a high-cost LNG tranche could not be rolled into the overall weighted average gas cost, the increment justifiable in terms of the implied improvement in security of supply.[13]

Owing to the liberalisation of natural gas markets in virtually all OECD regions, this is no longer true, while the resulting negative impact on project economics is further compounded by the fact that purchase agreements now recognise a buyer's right to reduce volumes to below contract levels, the so-called right of downward flexibility. True, this option is usually restricted with respect to both the time period over which takes may be reduced or the frequency with which purchasers may exercise this option. Even more significant is that having exercised the right, purchasers are free to sub-stitute, from the same project, lower cost spot for higher-priced contract supplies, thus lowering unit revenues again to the detriment of project cash flow. While take-or-pay contracts are obviously superior to conditional purchase commitments and, even more so, commodity sales, the terms and conditions governing recent agreements significantly reduce the implied revenue guarantee, both absolutely and compared with historic structures.

There are a few other organisational issues that need to be addressed as a prelude to our discussion of the leading features of the loan agreement itself. The two most important are ownership and organisational structure. Four ownership structures are commonly used to develop projects: incorporation, trusts, partnerships and unicorporated joint ventures. It could be argued that the choice of ownership structure is ultimately dependent upon the specific advantages, above all, tax benefits, connected with each structure. However, the evidence suggests there is little or no relationship linking ownership structure and putative tax benefits. This finding should not be too surprising, given that each of the four commonly used ownership structures confers, albeit to a different degree, tax benefits.

Buckley (1996b) delineates and schematically illustrates the interconnec-tions that exist within the four main project finance borrowing structures. The first of these is a project subsidiary. Under this structure, the borrower is a company owned, but not necessarily guaranteed, by the project sponsors. Borrower assets are confined to those used by the project; any assets owned by the sponsors but dedicated to project use are similarly charged to the bank. Project sponsors provide completion guarantees or, if a creditworthy entity, then directly by the contractor. The second structure, direct sponsor borrowing, lender recourse to the industrial partners is limited upon completion; lending banks now look mainly or solely to project assets to repay the loan. Such vehicles normally entail creation of an external escrow account, so that lenders can collect and control project revenues, ensuring that debt service obligations are met directly. One of the factors inhibiting widespread use of this form of borrowing vehicle is that the financing is arranged on behalf of project sponsors which lend on the funds. Project assets and liabilities, accordingly, have to be shown on the sponsors's balance sheet, with a potential deleterious impact on corporate leverage ratios.

The third financing option is the joint venture finance-operating company. Here sponsors enter into a separate agreement with the finance vehicle company, which in turn is vested with full responsibility for all phases of project operation and management including raising project debt, servicing the debt on the sponsor's behalf, paying expenses, remitting dividends and so forth. As with other limited recourse structures, sponsors provide guarantees both in respect of project completion and the management of the finance vehicle company. 'Since such structures normally entail a clear division of responsibility among sponsors, with respect to borrowing and/or marketing arrangements, their main intent is to preserve the joint venture for tax purposes, while differentiating each sponsors terms and conditions' (Buckley, 1996b).

The final option is leveraged leasing, which involves the lending banks, a lessor and a lessee as principals to the transaction. The bulk of funds are borrowed by the lessor, which retains all of the tax advantages connected with owning the leased assets. These advantages can then be passed on to the lessee through reduced lease payments; the lessee benefits since such payments, too, are tax deductible. On the other hand, since the assets are leased and not owned, they may not be available to the lender as security for the financing.

Determination of the optimal ownership structure, like the very issue of why sponsors choose limited recourse over lower cost balance sheet loans, is complex and only imperfectly understood. Traditional analyses seek to identify the way or ways project finance is able to reconcile the leverage, risk and tax interests of leading project participants. The available empirical evidence appears to support many, if not all, of the contentions raised in the theoretical literature. We argued previously that the traditional approach is not the last word in explaining either sponsor preference for or the unique characteristics of project finance debt. We shall not repeat why the options-theoretic approach seems to provide superior insights into all these issues. What does bear repeating is that the ultimate resolution of organisational and ownership issues depends upon the interaction of numerous influences; there are as yet no definitive answers to such questions.

Turning now to the loan agreement, the first critical issue to be addressed from a lender's standpoint is governing law. Most banks, quite obviously, prefer the application of their own law for any disputes that might arise under the loan agreement. Harries (1990), however, has argued that this principle does not apply with the same logic to project financings, since the choice of local law could in some instances be more of a burden than a benefit to the lending institutions. The favoured principle, according to Harries, should be to select as governing law the system most suitable to enforcement of the agreement or to a bankruptcy proceeding involving the project company. A second argument supporting the contention is that loan syndications for major international projects involve banks of different nationalities. For both reasons, the ideal system might be expected to have a well-developed commercial and financial code capable of addressing a wide range of complex commercial and legal issues; the law of England or New York would appear to meet all of these requirements, which explains why either is preferred for most international loan transactions.

The basic instruments that secure a creditor's interest in a project financing are mortgages, pledges and other encumbrances imposed upon the project company by the lending banks. Each of these instruments protects a secured lender against any other creditor seeking to enforce rights against certain project assets. When a project becomes insolvent, a secured lender may not always be able to sell off project assets to recover the remaining loan balance. Under these circumstances, lenders have two basic options: either to continue to operate the project by employing new project managers or, alternatively, to sell the project outright; legal opinion seems to favour the sale of pledged shares in the insolvent company rather than the project assets (Clifford Chance, 1991).

In the more traditional project financings, where the project company is located in an emerging market economy and the bulk of production is destined for export, lenders would normally insist upon the creation of an external escrow account through which the banks could exercise close and direct control over project revenues. Under this arrangement, export proceeds are deposited in the escrow account. Lending banks first disburse funds in an amount sufficient to cover debt service payments. Unless there are contingency reserve obligations, which have a prior claim on any residual sums, any remaining amounts can be disbursed to the project company.

The fourth and perhaps most critical feature of the loan agreement is the detailed completion guarantees provided by project sponsors as insurance against any mechanical or economic failure that would impair the project's ability to service its financial obligations to the banks. We have already indicated the nature of these undertakings. For present purposes it is sufficient to reiterate that these obligations are formally discharged through guarantees from the sponsors that they will cover outstanding loan payments until the project is constructed and operating up to speed. Such undertakings may require that the sponsor or sponsors complete the project to the technical standards contained in the project report, provide financing to cover investment cost overruns or, if needed, to infuse additional capital into the project to achieve completion, and ensure that all financial tests are met or exceeded upon completion.

As we have seen, in many instances sponsors are prepared to enter into sales and purchase agreements with the project that cover all or most project output and may, but need not, involve a concurrent fixed-price obligation. Depending upon the nature of the agreement, these commit-ments could be viewed as providing unconditional financial support to the project. In law such undertakings are not guarantees *per se*, but rather quasi-guarantees: they exist to underwrite cash flow, so that the greater the degree of unconditionality, the stronger is the implied support for the project. The strongest commitments of all will transcend *force majeure* or any other contingency that might interfere with the sponsors fulfilling their obligations under the contract. As a matter of fact, the project company will normally assign its rights under the sales and purchase agreement to the secured creditors.

Loan agreements also contain various financial covenants, the purposes of which are to protect creditors by ensuring that the sponsors always act with financial responsibility; the element of moral hazard that we mentioned at the beginning of this chapter is effectively addressed within the frame-work of these covenants. The sorts of obligations that are normally imposed include maintenance of an agreed debt-to-equity ratio during the life of the loan; restrictions on the distribution of dividends; a pre-set current ratio with a parallel obligation to satisfy working capital requirements; affirma-tive or negative pledges; events of default; and conditions precedent, which require that all contractual arrangements are in place, and that all arrange-ments and authorisations necessary for construction, operation and meeting debt service obligations are enforceable.

The last set of provisions worth considering are intercreditor relations, that is, agreements defining precisely the ranking of different project creditors. In virtually all project financings, lending banks will insist upon being senior to all other creditors. Even so, since disputes may arise among different creditors occupying the same creditor rank, there is the necessity of establishing clear rules as to how financial risks are to be allocated. Such agreements will also stipulate the ways in which the possibly divergent interests among lenders of the same rank can be protected and harmonised. This last consideration is especially important since it is not too difficult to imagine lenders falling out over conflicting interpretations of specific issues that might arise over the life of the loan.

The contractual framework is concerned with how individual contingen-cies will be addressed should they arise. Whether the final set of agreements provides the appropriate degree of protection is another matter, since the contractual framework can never be any better than the lenders's ability to identify, quantify and allocate project risks. Risk management is the hall-mark of leading project finance banks, and is achieved through, among other things, the employment of technical service experts, engineers and economists, who are able to judge on all technical and commercial aspects of the project. The 'other things' just alluded to include those practices used by banks to monitor and manage project risks including: conditions precedent; regular and complete reporting; interim and audited project financial statements; repetition of warranties and representations; drawdown restrictions and coverage ratios. At a more general level, there are four variables that are critical to successful risk management; each has already been covered in other contexts, so all that needs to be done here is to repeat their individual and collective importance and to show how they interact to determine the ultimate shape of the financing.

Cash flow lies at the core of limited recourse lending; it is worthwhile to reiterate the key factors affecting a project's cash flow potential and review the ways they are addressed by lenders. The most obvious is the type of business the project under consideration is engaged in. It goes without saying that lenders must have a clear understanding of the market for project output: the current strength of the market and how it might be expected to develop in the future. Market prospects encompass not only the outlook for demand but also the expected evolution of supply, above all, how the investment being considered for financing compares with alternative projects with respect to variables such as cost, location and so forth. The second critical element in the lending process is to test the robustness of the investment's cash flow potential to different states of the more important project variables. The importance of individual project parameters is established by sensitivity analysis and, as noted previously, the test should be applied multivariately, that is, assessing what impact changes in two or more key project variables might be expected to have on cash flow potential.

Other critical elements that need sorting out are whether sales are covered under long-term or spot contracts, whether purchasers of project output are creditworthy and the extent, where appropriate, that political, economic and currency risks have been catered for. All three points are straightforward and hardly require much elaboration. The first and second points are obviously interconnected. If all or the bulk of project output is sold to a single buyer, lenders need to be confident the purchase commitment will be honoured over the life of the financing. Accordingly, the value that lenders are prepared to impute to individual sales and purchase agreements varies directly with the buyer's creditworthiness. On the other hand, term marketing agreements are generally preferred to spot sales, although why is unclear. To the extent that contract sales involve offtake only, and thus involve the same degree of price risk as spot sales, the risk profiles of the two options with respect to cash flow potential appear identical.

The second critical feature of project financings is that lenders have access to project assets in the event of insolvency. The security inherent in project assets obviously depends upon several broad considerations, the most significant being whether they are dedicated solely to project use, as they should be in a classic limited recourse structure. The market value of project assets, and whether, in fact, it can be realised should the need arise, are closely related. Lending banks also need to be reassured that the project will always be properly maintained and, to that end, will need to know how such responsibilities are allocated and to whom. The importance of this point cannot be emphasised enough since operational efficiency is one of the main pillars upon which cash flow generation depends. Finally, it is imperative that the project has adequate insurance cover and that, to the maximum extent possible, all risks have been properly identified and assessed. This should be confirmed ideally by an independent insurance consultant with expertise in evaluating similar projects.

Apart from a project's cash generating potential, the second layer of lender security lies in being able to get hold of project assets and sell them, should the need arise, with the proceeds applied against any amounts outstanding under the loan. Analysts identify two states, aggressive and defensive, of lender security interests in project assets (Wood and Vintner, 1992). The aggressive aspect is manifest in lender insistence that rights over project assets must be conceded by the project sponsors as a precondition for obtaining financing. Project assets should be freely marketable, with a known price, while lenders must be able to exercise their rights without the need for any third party consents. Although conceptually sound, these principles are utterly unrealistic. In many instances there are no markets for project assets or, if they do exist, they are extremely thin, meaning that it is impossible practically to ascribe a fair market value to them. For partially completed projects, determination of asset values is even more questionable, the results of such attempts being at best arbitrary and at worst worthless. And finally, for emerging market projects, enforcement is almost always impossible without the host government's assent.

For all these reasons, there is a general consensus that the importance of security lies more in the legal advantages it confers on those having a charge over project assets than upon banks being able to seize and dispose of them (Clifford Chance, 1991; Wood and Vintner, 1992; Gewirtz, 1995). More to the point, a creditor with security over project assets ranks ahead of unsecured creditors, meaning that the ability of the latter to interfere in the relationship between the debtor and secured creditors is limited. Moreover, depending upon the prevailing legal system, a security interest may entitle lenders to use project assets as opposed to having the right to dispose of them.

The importance of insurance has already been noted, and lenders will require that the more significant project risks are covered with comprehensive policies. Comprehensive in this sense refers not only to amount but also to the nature of the risks covered and exclusions and deductibles, if any. Furthermore, lenders will want to ensure that their interests in the coverage are adequately protected through, for example, co-insurance, in effect becoming direct beneficiaries of separate and independent policies. A comprehensive insurance programme for a typical project would encompass the types of coverage shown in Table 5.3. As the table makes clear, capacities for the relevant insurance vary considerably. Not surprisingly,

Table 5.3 Types of insurance coverage and capacity

Type	Against	Capacity
All risk	Physical loss	Large
Design risk	Poor design	Small
Loss of profits	Consequential loss	Nil/small
Political	Expropriation, war, transfer	Medium/large
Insolvency	Third-party credit standing	Nil/small
Shortfall	Volume reduction	Nil

the larger the available capacity the lower will be the associated premiums. Insurance capacity to underwrite financial viability tends to be scarce and accordingly expensive. For most other major risk categories, capacity seems adequate.

Despite best efforts to the contrary, projects can and do fail, far more often than might have been expected. Nor does there appear to be any direct correspondence between the incidence of failure and the number of projects that are abandoned by their sponsors in favour of the lending banks. The lack of correlation is all the more remarkable given, as we shall see, that among the leading sources of project failure are poor cash flow and risk analysis. These points are confirmed by Castle's (1986) review of North Sea projects and Mikesell and Whitney's (1987) analysis of the global mining industry. Since these results are confined to the natural resource sector, it is arguable whether they can be taken as representative of the general tendency among other industrial sectors. Nor do the authors attempt to link project failure with abandonment, thus shedding no light whatever on the connections that might exist between the two.

One plausible explanation for the significantly higher incidence of project failure than abandonment is that mining or natural resource investments generally, are viewed by their sponsors as having options-like characteristics and accordingly valued in keeping with traditional options pricing models.[14] Within an options-theoretic framework high mineral price (and asset) volatilities and long project lives imply considerable upside potential. What really matters, therefore, is the sponsor's *ex ante* evaluation of project returns, which although conditioned by, is independent of, actual performance. Investors stick with marginal projects because of the implied sizable expected payoff, in spite of an abundance of empirical evidence suggesting project investments are worth less than the implied options valuations. Sponsors, in other words, appear to pay too much for natural resource projects – and by implication lenders must also be providing too much project debt; lenders, however, are implicitly compensated by the strength of investor optimism with respect to the project's eventual pay-off.[15]

Returning to the Mikesell and Whitney study, their basic methodology is to classify mining project performance into one of three categories based upon *ex post* financial results. Projects having an internal rate of return (IRR) in excess of 15 per cent are deemed to have enjoyed 'excellent' success. 'Successful' operations are defined as those where revenues exceeded cash operating costs, while projects are said to have failed if costs exceeded revenues. The test was applied to mining projects initiated between 1970 and the mid-1980s. Of the total number of mines surveyed, 20 per cent had excellent success as defined above, 44 per cent were successful while 36 per cent were failures. In other words, only one in five mining developments over the sample period met or exceeded investor expectations. It is even more discouraging that, when the revenues of all the projects surveyed are cumulated, they failed to cover combined operating costs.

Mikesell and Whitney further identify the leading types of problems experienced by mining projects included in the sample that failed, together with an indication of how widespread these problems were. In several instances more than one problem contributed to project failure. It is useful for present purposes to differentiate the importance of financial failure from poor design or operational failure. Among the categories that relate to financial failure, nearly 30 per cent were plagued by delays and cost overruns in the construction phase, while one-third incorrectly assessed prospective market prices for project output. Among other reasons, the single most important source of technical failure was in the metallurgical plant, closely followed by similar difficulties in milling operations. Finally, nearly one in four projects suffered from poor management.

The lessons Mikesell and Whitney draw from this experience is that 'companies most conscientious about planning and risk analysis have been the most successful and this generalisation is true regardless of size. Successful companies generally apply cash flow and risk analysis beginning in the early stages of mine development.' Even though it is not possible to extrapolate what the authors have to say about the mining industry to other natural resource sectors the overall conclusion, that project success is largely a function of careful planning and risk analysis, does seem broadly applicable, and corresponds closely to the central arguments of this study.

Notes

1 Sponsors will in most instances have recourse to the licenser, meaning that within the framework of a project finance loan they have direct, but not ultimate, responsibility for technical failure.
2 Several studies suggest that while resource depletion or scarcity cannot explain the pattern of oil or natural resource price developments over the past two decades, the importance of depletion, and hence its relation to prospective price developments, does differ between petroleum and natural gas on the one hand

and industrial metals on the other. For example, the ratio of user cost to market price averages 0.4–0.5 for petroleum resources as against 0.05–0.2 for bauxite, nickel and copper; these estimates are derived from Mueller (1985) and Stollery (1983). User cost is defined as the difference between price and marginal production cost. These differences appear to provide some justification for differentiating the long-term price outlook for oil and natural gas from that of major metals. On the other hand, they are hardly so great as to provide a compelling reason for doing so.

3   The stability aspect of transfer pricing derives from the fact that prices are typically set as a moving average of past, usually market, prices to avoid any obvious economic or financial distortions. The smoothing feature of transfer pricing implies lower volatility compared with market-sensitive pricing.

4   Brent eventually established itself as an alternative to, significantly, Saudi Arabian light, which prior to 1986 was the principal marker crude, that is, the price the OPEC cartel sought to, and for some time succeeded, in administering by agreement among member countries to proration production.

5   An equally interesting structure was applied to the financing of the Ekofisk project located in the Norwegian sector of the North Sea. A security package based upon the concept of a throughput agreement was developed to allow consortium members to achieve the desired degree of leverage. More to the point, the six industrial partners entered into a throughput and deficiency agreement with each of the four transport and processing companies. The parent company of each partner was also a partner to the throughput agreement. The four borrowing companies then assigned their rights under the agreement to a common trustee for the benefit of the lenders. Banks were thus highly secured, having recourse to the project, the industrial partners and the parent oil companies (Cox, 1983).

6   It could be argued that comparisons between the North Sea and the former Soviet Union are hardly apposite. Russia has only limited experience of having to deal with foreign investors, is politically unstable while there is some question as to how strong the commitment to a market economy really is. However, it is worth recalling that North Sea development occurred under the aegis of an (old) Labour government, with an equally uncertain commitment to a liberal market economy, while Tony Benn, then as now on the extreme left wing of the Labour party, was Minister of State for Energy for most of the region's early development.

7   Post-completion borrowers can still lose limited recourse status if breaches arise that are regarded as being within the borrowers' control. Failing to develop or operate the project in keeping with good commercial practice constitutes an example of such a breach. If, by contrast, the breach arose from, say, the arbitrary revocation of a licence provided by a government, there would appear to be no basis for withdrawing limited recourse status; in effect, the risk would have been assumed by the lending banks (Clifford Chance, 1991).

8   One of the more contentious issues is how the steam is priced. If purchase costs are linked to the quantity of gas used, there will be an obvious incentive to economise on the steam input; on any other terms, purchasers will, more likely than not, use the steam inefficiently.

9   The debate going on within Pakistan has already turned nasty, with the Nawaz government insisting power rates are exorbitant and therefore need to be reduced drastically. The Central Bank of Pakistan recently ordered banks to comply with an interim court order banning HubCo, one of the first independent power projects to be constructed, from transferring funds abroad. Some-

what incongruously, HubCo is said to be secure from further official attack, given extensive World Bank involvement in the project. Other foreign IPPs, too, are thought to be equally 'bullet proof', given that these agreements require international arbitration in the event of dispute. Unilateral cancellation of IPP projects could trigger penalties ranging from $6bn–$12bn, not to mention various other forms of international financial retaliation, including freezes on official concessional finance (Power in Asia, 1998).

10   According to recent surveys, east Asian power equipment orders are projected to fall sharply over the next five years. According to the figures given in the *Financial Times* (Marsh, 1998), order volumes for gas turbines are projected to decline over 1998–2002 compared with the preceding four year period by between 30 per cent (Japan) and 50 per cent (South Korea). For the region as a whole, the direct effect of the crisis will be to cut some 7–8GW a year out of the overall market. Another casualty of the Far Eastern crisis is that Qatar's Ras Laffan bonds, the first instance of a major international project financing including a bond tranche, were recently downgraded owing, *inter alia*, to the less favourable regional economic and financial outlook and changes to the structure of the original sales and purchase agreement, above all, the recent elimination of a guaranteed price floor for project output.

11   The conviction still remains strong among some analysts that take-or-pay contracts are the *sine qua non* for financing large mineral, above all, European natural gas, projects (Stern, 1992). Walde (1993), on the other hand, maintains that no one can any longer take such arguments seriously, whether in respect of energy (oil, natural or liquefied gas, coal or uranium) or other natural resource industries. In all these cases, volatile spot and contract markets have replaced long-term contracts with captive consumers without any perceptible impact on the willingness of commercial lenders to provide the required finance. Even so, opponents of market liberalisation allege that the European gas market is unique, and therefore comparisons with other regional gas markets or mineral industries are misleading. Walde rightly concludes that the 'burden of proof is on the opponents of a more liberalised access system to gas pipelines to prove persuasively that the gas industry is really – for technical/economic reasons or otherwise – different from other energy and commodity industries... This will be difficult to prove.'

12   The financing of major international LNG projects is recent phenomenon, owing to the complex interactions that exist among the upstream, downstream and transport phases of such projects, huge capital costs – again across all project phases – political risk factors and so forth. See Woicke (1983) and Koiishi (1983) for a discussion of the relevant issues in the early days of industry development. The current state of the project finance market as it relates to LNG is given in *Petroleum Finance* (1997).

13   An alternative explanation for the willingness of gas buyers to enter into high cost supply contracts is provided by Leslie and Michaels (1997): 'natural gas markets are local and opaque because of the difficulty of storing and transporting gas. The greater the uncertainty over future investment (and therefore production) plans, the greater the price volatility; and the greater the incentive for gas buyers to commit to high priced supplies.' Ironically, liberalisation of the European gas market has led to behaviour opposite to that predicted by Leslie and Michaels. Accordingly, we must either assume economic irrationality on the part of gas buyers or alternatively accept the possibility that the underlying assumptions made by the authors are false. The bulk of available evidence argues strongly in favour of the latter interpretation.

14 'In many industries, companies stay in business and absorb large operating losses for long periods, even though a conventional NPV analysis would indicate that it makes sense to close down the factory or go out of business... Closing down a plant would have meant an irreversible loss of tangible and intangible capital... Continuing to operate keeps the capital intact and pre-serves the option to resume profitable operations later. The option is valuable, and therefore, companies may quite rationally choose to retain it, even at the cost of losing money in the meantime' (Dixit and Pindyck, 1995).

15 An alternative explanation is that sponsors are extremely reluctant to default on project loans for fear their credit rating may suffer. This conclusion is based on the well known principle that banks 'forgive their enemies, but never forget their names'. The problem with this explanation is it ignores the fact that sponsors chose, and paid a premium for obtaining, limited recourse status. Are we to assume this was done without sponsors ever expecting to exercise this option, except perhaps in the direst circumstances imaginable.

6 Four Case Studies

This chapter presents a number of case studies that illustrate the main concepts and themes discussed in previous chapters. Even the simplest of projects are invariably complex. No purpose, therefore, would be served by trying to capture all of the nuances in the four cases we shall be reviewing. Indeed, the analysis has been kept as simple as possible to highlight the critical issue each was selected to address.

The first case study introduces cash flow analysis within the framework of a simplified oil financing in the United States. In previous chapters it was noted that a project unlikely to cover its required rate of return can never be of interest to commercial lenders. The case is designed to show how, once it has been established that an investment is economically viable, lenders determine a project's maximum loan value. No attempt is made to derive credit benchmarks that are critical inputs into such a determination. The analysis is instead developed with the coverage ratio taken as given; coverage ratios for energy projects are well defined and accordingly do not require further explanation (Pollio, 1986, 1992). An equally significant aspect of the case is to show how commercial lenders use sensitivity analysis to refine project loan values. From an investor's standpoint, the crucial question is how changes in the behaviour of key variables will affect the project's NPV. For commercial lenders, the benchmark is the sensitivity of loan coverage ratios to different states of the principal project variables.

The second case study focuses on the alleged shortcomings of traditional discounted cash flow methodologies, in particular, their inability to capture the value inherent in being able to manage project assets flexibly. The case is based upon a highly stylised analysis of alternative power plant investment options. Indeed, the study upon which the present case is based was deliberately contrived to show how recognition of real project options can fundamentally alter conclusions about the economic viability of an investment based upon traditional methodologies. The study shows that an economically *unviable* investment, as measured by traditional discounted cash flow analysis, actually emerges as the dominant option once managerial flexibility is properly accounted for.

Far more instructive is what is omitted from the original analysis, namely, a failure to show how real options can be engineered within existing project structures. This omission is all the more unfortunate in that one of the main virtues of the real options approach over more traditional methodologies is precisely the fact that it recognises and rewards such flexibility. Once this adjustment is factored into the analysis the authors' preferred investment option ceases to be optimal. Indeed, our results indicate the preferred project is now the one that offers investors the greatest flexibility at the lowest investment cost affirming, albeit by a different route, the main point of the original study. On the other hand, our findings undercut one of the central premises of the initial case, namely, the mutual exclusivity of the two methodologies. They demonstrate that traditional discounted cash flow methods are adequate for evaluating most investment opportunities, once the analysis is appropriately formulated.

The third case study looks at the economic rationale for undertaking a liquefied natural gas project (LNG) in Qatar, a small oil exporting emirate in the Arabian peninsula. The background to the case was touched upon briefly in the first chapter. We noted there that Mobil, one of the venture's industrial investors, was experiencing reserve difficulties in Indonesia, where the bulk of its LNG investments were located. The choice of Qatar, the site of two projects Mobil had decided it would develop concurrently, was not immediately obvious. True, the logical disposition of project output was in the Far East, the destination of the bulk of Mobil's existing LNG production, and where the company's LNG marketing capabilities were presumably the strongest. On the other hand, there were no overwhelming reasons favouring Qatar. Numerous other locations could have served the Asian market equally well, so the issue boils down to the comparative economics of locating two LNG projects in Qatar as opposed to somewhere else.

The Qatari projects are instructive for other reasons as well. Construction and financing of two huge back-to-back LNG projects in a small oil producing country posed potentially formidable portfolio saturation problems. One way round this difficulty was to try to access funds for project development in two distinct international markets, the traditional project finance loan market and the newer, but untested, fixed income market.

The logic of tapping the project bond market was unexceptionable. Ras Laffan possessed all the features common to traditional project financings. Export sales were denominated in hard currency and underpinned by strong contractual commitments, including a guaranteed minimum price, provided by a creditworthy Korean utility. And second, project management was in the hands of one of the world's ablest producers and marketers of LNG. For investors the attractions of the issue were obvious: it represented a more or less pure natural gas investment that tied into the growth of one of east Asia's largest and most successful economies. It is important to note that in the present context commodity investment does not mean that returns are linked in any way with the price of liquefied natural gas.[1] Ras Laffan bonds, rather, were underwritten by the expected strong growth of the Korean economy, with its attendant favourable impact on current and future gas demand. As we have seen, the factors shaping the bonds' initial appeal proved to be the primary source of their subsequent weakness.

The fourth and final case study, expansion of an existing fertiliser operation in Pakistan, lies at the exact opposite end of the spectrum.

Pakistan is currently unable to borrow on its own guarantee while the bulk of planned project output is intended for domestic consumption; only a very modest tranche of future production is ever likely to be available for export. The economic benefits of the project thus lie overwhelmingly in two areas: higher local agricultural productivity and import substitution with concomitant savings of foreign exchange. Accordingly, the investment is best viewed as an international development project. It will come as no surprise, therefore, that its financing involved major international development banks while the commercial portion of the financing was insulated from all political risks. Immediate interest in the project derives less from any lessons it may have to offer in respect of project finance – it has few – than from determination of an appropriate return to be applied by an international lending institution with an equity stake in the project.

One of the distinguishing features of many emerging market economies today is the willingness to equate social with private discount rates. This approach, representing a radical departure from past practice, is still far from being applied universally. Equivalence ultimately depends upon acceptance of the fact that the historic dichotomy between public and private rates of return was spurious. Since capital resources are scarce and have alternative uses, the private sector's required rate of return provides the soundest index of an investment's wealth-creating potential. This is not to say national investments will no longer be subsidised. Rather, it makes the issue of public subsidies a matter of political, not economic, debate. Such economic distinctions were thought to apply more appropriately to richer, economically advanced countries. Poorer nations like Pakistan, some would still argue, deserve special dispensation if their economies are to converge to those of their more prosperous neighbours. The fact that development agencies now insist upon returns commensurate with a project's 'true' opportunity cost shows how reluctant they are in the main to 'buy into' such arguments. It is essential, therefore, that required returns are calculated correctly. Otherwise, there is the risk of unnecessarily penalising a project if its hurdle rate is too high or providing an implicit subsidy if required returns are set too low.

AN OIL FINANCING IN THE UNITED STATES

Petroleum has been known and used for centuries, although only over the past 150 years has it been exploited commercially. The beginning of the modern petroleum industry is usually dated to 1859, when commercial production first commenced in Pennsylvania. The main historic end-use product was kerosene, which initially replaced whale oil for illumination. The United States was long the world's dominant petroleum producing nation; indeed, even as recently as the mid-1950s there was sufficient spare local production capacity to make good import losses sustained by Great Britain during the Suez War. Moreover, the United States was home to five of the seven largest integrated oil companies. These companies are diversified both geographically and functionally, that is, their operations encompass upstream (production) and downstream (refining, transportation, petrochemical production, distribution and retail) activities, although the relative importance of and the activities carried on within each sector vary considerably from company to company. Production of oil and gas occurs in a number of American states, with the southwestern part of the country accounting for the bulk of national output.

We have already seen that the acquisition of potential reserves can be conceptualised as being distinct from the decision to develop them; this is the central insight of the options-theoretic approach. Whether it makes economic sense to develop the resource is conditional upon prospective returns exceeding the investors' opportunity cost of capital. That is, whether the project is able to produce a positive NPV. The following example, adapted with modification from McKechnie (1990), illustrates the mechanics of this process. The example was originally intended to show how commercial banks set loan values and use sensitivity analysis to test the robustness of project cash flow to alternate states of key project variables. This aspect of the case has been retained.

Table 6.1 presents the project's base case economics. The cash flow model is based upon technical, economic and market parameters. Technical parameters include the size of the underlying reserve, the resulting production profile and the costs likely to be incurred to produce the resource. Each of these parameters will have been analysed in the project report from which the relevant line entries are taken. Economic variables include, above all, the rate of inflation and the rate of interest applicable to the project loan, while market variables include the price of crude. For analytical purposes, base case oil prices are set at $15 per barrel and projected to escalate by 5 per cent per annum from the third project year. Operating costs, by contrast, are assumed to rise at an average annual rate of 10 per cent immediately following commencement of project operations. The remaining data inputs are an assumed interest rate of 10 per cent, an income tax rate of 55 per cent and a four year repayment schedule payable in eight semi-annual installments.

The first line of the spreadsheet shows the volume of oil the project is expected to produce expressed in million barrels, a conventional industry metric. This is an engineering estimate developed by the sponsor's reservoir engineer and vetted by the lending bank's technical services unit.[2] Lending banks may accept the original reserve estimates unchanged; alternatively, they may evaluate project economics based upon more conservative reserve

Table 6.1 Base case cash flow model

	Y0.5	Y1	Y1.5	Y2	Y2.5	Y3	Y3.5	Y4	Y4.5	Y5	Y5.5	Y6	Y6.5	Y7	Y7.5	Y8	Y8.5	Y9	Y9.5	Y10
Oil production (MMBBL)	0	0	0	0	3	3	8	10	10	10	9	9	8	8	7	6	5	4	3	2
Oil price $/BBL	15	15	15	15	15	15	15.4	15.8	16.1	16.5	16.9	17.4	17.8	18.2	18.7	19.1	19.6	20.1	20.6	21.1
Gross revenue	**0**	**0**	**0**	**0**	**45**	**45**	**123**	**158**	**161**	**165**	**152**	**157**	**142**	**146**	**131**	**115**	**98**	**80**	**62**	**42**
Capital costs	60	75	100	95	10	10	0	0	0	0	0	0	0	0	0	0	0	0	0	0
Operating costs	0	0	0	0	15	16	16	17	18	19	20	21	22	23	24	25	27	28	29	31
Taxes	0	0	0	0	0	0	0	0	18	76	69	72	65	67	59	49	39	29	18	6
Net cash flow pre debt service	**(60)**	**(75)**	**(100)**	**(95)**	**20**	**19**	**107**	**141**	**125**	**70**	**63**	**64**	**55**	**56**	**48**	**41**	**32**	**23**	**15**	**5**
Initial loan balance	0	52	120	213	306	306	306	268	230	192	154	116	78	40	2	0	0	0	0	0
Loan drawdown	51	64	85	81	0	0	0	0	0	0	0	0	0	0	0	0	0	0	0	0
Loan repayment	0	0	0	0	0	0	38	38	38	38	38	38	38	38	2	0	0	0	0	0
Interest rate	1	4	8	12	15	15	14	12	10	8	7	5	3	2	0	0	0	0	0	0
End loan balance	52	120	213	306	306	306	268	230	192	154	116	78	40	2	0	0	0	0	0	0
Cash flow to sponsor	**(9)**	**(12)**	**(19)**	**(22)**	**5**	**4**	**55**	**91**	**77**	**24**	**18**	**21**	**14**	**17**	**48**	**41**	**32**	**23**	**15**	**5**
NPV @ 10% net cash flow pre debt service	250	326	421	547	674	686	701	623	506	401	347	298	246	200	152	109	72	42	20	5
Cover ratio (PV NCF/end loan balance)	4.81	2.72	1.98	1.79	2.21	2.25	2.62	2.72	2.64	2.61	3.00	3.85	6.23							

IRR to sponsor

Oil price $/BBL esc. @ (%)	5	Operating costs esc. @ (%)	10	Interest rate (%)	10
From (year)	3	Taxes (%)	55	Loan repayment equal s.a.	4

Source: McKechnie (1990).

Key: BBL – barrels
MMBBL – million barrels

assumptions. Note that during the first two years the project is not expected to produce any oil at all. Thereafter, production peaks soon after start-up, and declines continuously over remaining project life.[3]

Line 3 shows the oil prices used to generate project revenues. The figures given are the lending bank's price premises, which again may or may not correspond to those contained in the project report. Price premises are based only in part on fundamental analysis. The economic component is derived from an objective assessment of market dynamics, provided either by economists employed by the lending bank or outside industry consultants. Given the centrality of price forecasts to deriving a revenue baseline, major energy banks typically have an internal price premises committee vested with responsibility for developing oil and natural gas price premises and forecasting rates of price and cost escalation and other variables used in connection with the bank's evaluation of the project.

The term 'premises' is used in preference to forecasts to highlight the importance of non-economic factors in their derivation. Banks eager to engage in this type of lending will normally project a stronger rate of price escalation than the consensus rate of increase; even so, the variance in the price premises used by leading energy banks tends to be comparatively small. Finally, the principle followed in the case to hand is that oil prices are forecast to increase in nominal terms only and at the same rate as the forecast rate of inflation; real oil prices, in other words, are expected to remain unchanged over project life. This has long been banking industry practice.[4]

Line 4 presents base case gross revenues, the product of estimated oil production and forecast oil prices. Cash outflows are given in the following three lines. Line 5 shows estimated project investment costs, while line 6 gives project operating expenses. Production costs are escalated at twice the assumed rate of price increase. The corporate tax rate over project life is 55 per cent and is applied against gross revenue less capital and operating costs; tax losses are cumulated and carried forward indefinitely. Line 8 shows net income after taxes. In the construction phase, this line shows the amount of capital needed to complete the project; in the operating phase, the entries correspond to the amount of cash the project is expected to generate without regard to the way the project was financed.

The next four lines model the project loan. Banks are assumed to cover 85 per cent of the project's capital costs at a 10 per cent interest rate; interest is capitalised at the same rate during the construction period. Line 14 shows the amount of cash available for distribution to shareholders after debt service obligations have been met. Cash flow is negative during construction, reflecting the sponsor's equity contribution, that is, the remaining 15 per cent of project investment cost. The following line shows the present value of the project's remaining net cash flow (before debt service) and discounted at an assumed 10 per cent rate.

Line 15 presents the project's net present value, while the next line gives the cash flow coverage ratio, a concept that will be discussed in due course. The positive NPV shown in line 15 signifies the project is economic and worthwhile undertaking. This determination, however, is based upon the lender's, not the sponsor's, opportunity cost of capital. Since within the framework of a limited recourse financing, the borrower has a right of abandonment in favour of the lending banks should the project subsequently prove uneconomic, it is essential to determine whether the project is able to cover the investor's hurdle rate.

The capital asset pricing model described in Chapter 2 is one way of determining the investor's required rate of return. The critical inputs into the CAPM are beta, the risk-free rate of interest and the market premium. Since this is an upstream investment, the relevant beta should relate to oil and gas producers. The composite beta for US upstream companies is 0.92 (Merrill Lynch, 1997). However, we know one of the critical determinants of beta is the degree of leverage. The project to hand is highly leveraged, with a debt ratio more than twice the 40 per cent sector average. To determine the appropriate value of beta, it is necessary first to unlever the reported beta, and replace the average sector debt/equity ratio with the degree of project leverage. An unlevered beta is defined by

$$\beta_U = \beta_L(1 + [1 - \tau]D/E),$$

where β_U and β_L are, respectively, unlevered and levered betas, τ is the rate of corporate taxation and D/E is the debt to equity ratio. With β_L given as 0.92, τ equal to 55 per cent and D/E at 66.67 (corresponding to a 40 per cent debt to total capital ratio), the unlevered beta equals 0.86 corresponding to a project beta of 4.12. Now, if we assume a market risk premium and a risk-free rate of 6 per cent,[5] then the required equity return amounts to 30.7 per cent while the project's weighted average cost of capital is 13.1 per cent. Since the project's internal rate of return is 33.3 per cent, the investment as seen from today will easily have exceeded its hurdle rate.[6]

The next step is to determine the maximum loan value a lender would be prepared to ascribe to this project based upon the cash flow profile shown in Table 6.1. Finnerty (1996) presents a stylised cash flow model that is extremely useful for illustrating in general terms the broad range of factors affecting the scale of project loan commitments.[7]

To illustrate how Finnerty's model can be used we assume, for the sake of simplicity, that non-cash costs are zero and the growth rates of project revenues and operating costs are the same, that is, $g_R = g_E$, in which case the present value of project cash flow is given by

$$PV = (1 - t)(R - E)/i - g_R\left[1 - \{1 + g_R/1 + i\}^N\right].$$

The resulting present value, when divided into the lender's target coverage ratio, yields the project's maximum loan value, the maximum amount of debt the project is capable of bearing at a given interest rate over a given time period N. The ultimate size of the loan value is thus shown to be an increasing function of both g_R and N. A higher rate of revenue growth is consistent with an increased cash flow enabling the project, all things constant, to support a higher loan value. As the loan tenor increases, there will be a corresponding reduction in annual debt service payments again consistent with a larger loan value.

One important qualification worth noting is that the analysis assumes net revenues and loan disbursement occur immediately. In the more usual case, there will be a gap in timing between the two. The loan will be drawn to construct the project and only after completion will the investment begin to generate net revenue. Finnerty models this gap as $PV^* = PV/(e^{iM})$ where PV is as defined above. Since the project must meet the lender's coverage constraint (α) it follows that $PV = e^{iM}\alpha D^*$ so the maximum loan value now becomes $D^* = PV/[\alpha(e^{iM})]$. The maximum loan value is an inverse function both of the rate of interest and the length of time revenue generation is delayed.

We can test these general principles with respect to the oil project outlined in Table 6.1. More to the point, we shall first evaluate the maximum size loan the project can bear, given the project's cash flow profile and the typical bank coverage ratio. Next, we shall review what impact changes in key variables are likely to have on project cash flow and how this will affect the project loan value. To this end, four of the principal project variables, namely the scale of oil production, capital costs, interest rates and the price of oil, are varied to test the robustness of the coverage ratio to such changes. Note, too, that the sensitivities are presented on a pairwise basis – far more informative in our opinion than are one-way sensitivities – with each of the production and financial variables cross-referenced to the price of oil.

Table 6.2 presents the results. The first cell in each panel of the table contains the value of the variable as used in the base case. For example, the first cell of the first panel shows the coverage ratio corresponding to an oil price of $15 a barrel at 100 per cent of estimated production. Sensitivities are assessed at production levels as low as 50 per cent of the volume assumed in the base case, and capital cost overruns of up to 1.5 times the originally estimated amount. Interest costs fluctuate from 4 per cent below to 6 percentage points above the base case rate.

Consider first the impact of price changes on project economics. In the base case, projected discounted cash flow provides for a coverage ratio in excess of 2 to 1; even if prices were to decline to as low as $9 a barrel, there would still be enough cash flow to cover the loan amount outstanding. Project economics are shown to be even more sensitive to estimated production volumes: each 10 per cent reduction in estimated output results

Table 6.2   Sensitivity analysis of cover ratio

Base case oil price ($/bbl)	Percentage reduction in oil production												
	15	14.5	14.0	13.5	13.0	12.5	12.0	11.5	11.0	10.5	10.0	9.5	9.0
0%	2.11	2.05	1.98	1.92	1.85	1.79	1.73	1.66	1.60	1.53	1.46	1.39	1.32
5%	2.01	1.95	1.89	1.83	1.77	1.71	1.65	1.58	1.52	1.45	1.39	1.39	1.25
10%	1.91	1.85	1.80	1.74	1.68	1.62	1.56	1.50	1.44	1.38	1.31	1.25	1.18
15%	1.81	1.76	1.70	1.65	1.59	1.54	1.48	1.42	1.36	1.30	1.20	1.17	1.11
20%	1.71	1.66	1.61	1.55	1.50	1.45	1.39	1.33	1.28	1.22	1.16	1.10	1.04
25%	1.61	1.56	1.51	1.46	1.41	1.35	1.35	1.25	1.19	1.14	1.08	1.03	0.97
30%	1.51	1.46	1.41	1.36	1.31	1.26	1.21	1.16	1.11	1.06	1.00	0.98	0.89
35%	1.40	1.35	1.31	1.26	1.21	1.16	1.12	1.07	1.02	0.97	0.91	0.84	0.76
40%	1.28	1.24	1.19	1.15	1.10	1.06	1.01	0.97	0.92	0.87	0.79	0.71	0.63
45%	1.16	1.12	1.08	1.03	0.99	0.95	0.91	0.86	0.79	0.72	0.65	0.57	0.50
50%	1.03	0.99	0.96	0.92	0.88	0.83	0.77	0.70	0.63	0.57	0.50	0.44	0.37
	Percentage increase in capital costs												
0%	2.11	2.05	1.98	1.92	1.85	1.79	1.73	1.66	1.60	1.53	1.46	1.39	1.32
5%	2.04	1.98	1.91	1.85	1.79	1.73	1.67	1.61	1.54	1.48	1.41	1.34	1.27
10%	1.97	1.91	1.85	1.79	1.73	1.68	1.62	1.56	1.50	1.43	1.37	1.30	1.24
15%	1.91	1.85	1.79	1.74	1.68	1.63	1.57	1.51	1.45	1.39	1.33	1.27	1.20
20%	1.85	1.80	1.74	1.69	1.63	1.58	1.52	1.47	1.41	1.35	1.29	1.23	1.17
25%	1.80	1.74	1.69	1.64	1.59	1.54	1.48	1.43	1.37	1.32	1.26	1.20	1.14
30%	1.75	1.70	1.65	1.60	1.55	1.50	1.45	1.39	1.34	1.28	1.22	1.17	1.11
35%	1.70	1.65	1.61	1.56	1.51	1.45	1.41	1.36	1.30	1.25	1.20	1.14	1.08
40%	1.66	1.61	1.57	1.52	1.47	1.42	1.38	1.32	1.27	1.22	1.17	1.11	1.06
45%	1.62	1.58	1.53	1.48	1.44	1.39	1.34	1.29	1.24	1.19	1.14	1.09	1.03
50%	1.58	1.54	1.50	1.45	1.41	1.36	1.34	1.26	1.22	1.17	1.12	1.06	1.01
	Interest rate												
6%	2.13	2.07	2.00	1.93	1.87	1.80	1.73	1.67	1.60	1.53	1.46	1.39	1.32
7%	2.13	2.08	2.00	1.93	1.86	1.80	1.73	1.67	1.60	1.53	1.46	1.39	1.32
8%	2.12	2.06	1.99	1.93	1.86	1.80	1.73	1.66	1.60	1.53	1.46	1.39	1.32
9%	2.12	2.05	1.99	1.92	1.86	1.79	1.73	1.66	1.60	1.53	1.46	1.39	1.32
10%	2.11	2.05	1.99	1.92	1.85	1.79	1.73	1.66	1.60	1.53	1.46	1.39	1.32
11%	2.11	2.04	1.99	1.92	1.85	1.79	1.73	1.66	1.59	1.53	1.46	1.39	1.31
12%	2.10	2.04	1.98	1.91	1.85	1.79	1.72	1.66	1.59	1.53	1.45	1.38	1.31
13%	2.10	2.03	1.97	1.91	1.85	1.78	1.72	1.65	1.59	1.52	1.45	1.38	1.31
14%	2.09	2.03	1.97	1.91	1.84	1.78	1.72	1.65	1.58	1.51	1.44	1.37	1.30
15%	2.08	2.03	1.96	1.90	1.84	1.78	1.71	1.64	1.58	1.51	1.44	1.37	1.30
16%	2.08	2.02	1.96	1.90	1.84	1.77	1.70	1.64	1.57	1.50	1.44	1.37	1.30

Source: McKechnie (1990).

in a more or less proportional decline in the coverage ratio. Coverage ratios are marginally less sensitive to capital cost overruns, and exhibit virtually no response at all to changes in assumed interest costs. Even at the extremes of the analysis coverage ratios are only some 2–6 percentage points lower than in the base case.

The interactions that exist among the individual variables are far more interesting. If oil prices were to decline to the low case shown in the chart and output amounted to one half of base case volumes, the coverage ratio would be only marginally higher than zero. At the extremes of the price and capital cost sensitivities, by contrast, the coverage ratio falls to around 1. In the typical oil financing, the maximum loan value would be set so that, at all times, cash flow coverage under the base case scenario would be two or more times the outstanding loan amount *and* equal the outstanding loan value at the lender's low price case. If we equate $15 a barrel with the lender's base price, not unreasonable even today, and $10 with the low price case, then the project could easily support the existing loan value. Over most states of the key project variables, the coverage ratio is generally above 1, the only exception applying to a significant reduction of production with or without a concomitant decline in prices.

The same data can be used to illustrate the impact leverage has on the value of project equity. In Chapter 4 we argued that, within an options-theoretic framework, leverage can enhance equity value by increasing asset volatilities. The amount of debt used to finance a given project is likely to be far higher in a limited recourse than a straight loan. We noted above the composite debt to equity ratio of upstream US oil and gas producing companies was 40 per cent, roughly one half the leverage ratio common in limited recourse financed projects.

To give some idea of the significance of risk shifting, as this process is called, we have tabulated the equity value of the project given in Table 6.1 under different assumed volatility (σ) and time to expiration (t) assumptions. The calculations needed to derive a precise valuation are well beyond the scope of this study. For example, valuation of the call option requires as inputs the value of the options remaining after each contractual debt payment has been made plus the value of the entire 16 options (Brealey and Meyers, 1991). The objective of the present discussion is to illustrate how project finance is used to create value for equity holders, so the issue of valuing nested options is sidestepped.

The first point worth noting is that options valuation methods do not provide an estimate of project value (P_V), only the way asset values are distributed between project equity (P_E) and debt (P_D) holders. Second, for present purposes we use the face value of project debt (D) and equate the time to expiration with the loan tenor, 4 years in the case to hand. Finally, project value is assumed equal to the investment cost of the project as given in Table 6.1. Since no output occurs during the two year construction period, the opportunity cost of the development lag equals the present value of lost production revenue, discounted at the investment's assumed dividend yield. Dividend yield is defined as net production revenue divided into reserve value. Paddock *et al.* (1988) indicate that 5 per cent is a reasonable approximation, and that is the value used here. Finally, the valuations shown in Table 6.3 were derived applying the Black–Scholes call option solution to these data inputs.

Three different volatility and time to expiration assumptions were used to generate the values shown in the table. Asset volatilities range from 40 per cent to 60 per cent annually. There is, of course, no one to one mapping between debt to equity ratios and volatility estimates so it would be wrong to equate a 40 per cent leverage ratio with the equivalent asset variance. Still, it should be clear that the volatility of asset values will be higher for limited recourse debt than for balance sheet loans. We have also evaluated option values for tenors ranging from 4 years, as in the example to hand, to 8 years. In the current market, average tenors are generally longer than the maximum shown in the table, suggesting that the degree of risk shifting inherent in project finance loans may be considerably greater than the

Table 6.3 Black–Scholes estimates of project equity and debt

A. Black–Scholes equity model

Project value $P_V = P_E + P_D$
Project equity $P_E = P_V N(d_1) - D/e^{rt} N(d_2)$
Project debt $P_D = P_V - P_E$

B. Value of project equity and debt for different volatility and loan tenor assumptions

σ	P_E			P_D		
	$t=4$	$t=6$	$t=8$	$t=4$	$t=6$	$t=8$
0.20	122.5	143.6	160.3	227.5	206.4	189.7
0.40	159.6	184.5	204.6	190.4	165.5	145.4
0.60	196.6	225.9	266.8	153.4	124.1	83.2

C. Probability of default on debt

σ	$t=4$	$t=6$	$t=8$
0.20	0.36	0.42	0.46
0.40	0.55	0.61	0.64
0.60	0.66	0.72	0.86

Note: $d_1 = \ln(P^*/D) + (r + 0.5\sigma^2)t/\sigma\sqrt{t}$, $d_2 = d_1 - \sigma\sqrt{t}$, $P^* = P_V/e^{yt}$. All the variables are as defined above and D is the face value of project debt, t the loan tenor, y is the dividend yield and r is the risk-free rate, assumed equal to 5 per cent.

simulations indicate. Project bonds would appear to offer sponsors an even greater potential to boost equity values, given that tenors more closely approximate to project life and thus are considerably longer than those applying to project loans.

If we assume for analytical purposes that annual asset volatilities consistent with the average industry debt to equity ratio and with the debt to equity ratios common in limited recourse loans are 0.4 and 0.6, respectively, then, at the indicated 4 year tenor, the latter will have increased equity value by about one-quarter compared with traditional debt ratios. If loan tenors are increased to 8 years, and the higher volatilities continue to apply, the value of equity will have risen by additional 5 percentage points. At the extremes shown in Panel B, the differences in equity value exceed 100 per cent.

The converse is, of course, a corresponding reduction in the value of project debt. With loan tenors given at 4 years, the market value of project debt at 40 per cent volatility amounts to two-thirds of its face value, declining to just over one half at the 60 per cent volatility level. The lower the ratio of market to face value, the higher will be the expected probability of default on project debt, as shown in the last panel of the table. In the Black–Scholes call option solution, the value of project debt is shown to be a function of the *adjusted* discounted present value of the debt, with the adjustment factor, $N(d_2)$, the probability that the equity position will finish in the money. Accordingly, $1 - N(d_2)$ is the probability project debt will default. The last panel of the table shows default probabilities corresponding to different volatility and loan tenor assumptions. At the extremes of the table, the probability of default varies by a factor of more than 2.5.[8]

To summarise, these simulations demonstrate the zero sum character of the process and why limited recourse debt is an ideal vehicle for addressing such risks. Sponsors are effectively barred from deploying project resources for any purposes other than those spelt out in the loan agreement. Also, since project assets cannot be commingled they are more transparent, which of course simplifies monitoring. And finally, within the framework of traditional project finance loan structures, creditors typically collect and control the disposition of project revenues.

EVALUATING MUTUALLY EXCLUSIVE POWER INVESTMENTS

The present case is intended to illustrate the potential shortcomings of traditional discounted cash flow methodologies. The analysis is derived from a study by Kulatilaka and Marcus (1992), which looks at the economics of three mutually exclusive power projects being considered by an electric utility. The issue to be decided is the choice of fuel input, namely, whether the utility should commit either to heavy fuel oil or natural gas, or opt instead for a facility capable of using both fuels. In all cases, project life is assumed to be 10 years. Gas prices are forecast to remain stable at $1 per million British thermal units (mmbtu) over project life. Oil prices are currently assumed equal to $0.75/mmbtu, but are projected to rise over the next decade at 5 per cent per annum. Regardless of which technology option is chosen, revenues are equal to $1.16 million in the first year of project operation, and increase thereafter at 5 per cent annually.

In the first 5 years of project life oil enjoys a clear cost advantage over natural gas. On the other hand, gas becomes more cost-effective over the remainder of project life. The capital costs of constructing each single fuel facility are the same, namely, $2.5 million. The utility has a third choice that would allow it to use either hydrocarbon as a fuel input. However, the capital costs of the dual-fuel project, at $3.0 million, are 25 per cent higher than either of the dedicated investment options. More positively, the operating costs of the dual-fuel boiler are the lower of oil or natural gas prices, since the boiler has been designed to use either fuel.

Spreadsheet analysis of the economics of the three projects is given in Table 6.4. Net present values associated with each option are summarised below, with the cash flows discounted at 10 per cent (all figures are in millions of dollars):

$$NPV_{OIL} = -\$2.5 + \$2.73 = \$0.23$$
$$NPV_{GAS} = -\$2.5 + \$2.45 = -\$0.05$$
$$NPV_{DUAL} = -\$3.0 + \$2.97 = -\$0.03$$

Applying traditional discounted cash flow analysis, only the oil-fired option produces a positive NPV and thus would be the selected investment. The alternative investments fail to cover the assumed 10 per cent cost of capital and accordingly would be rejected. Note that the dual-fuel option produces a higher payoff than does natural gas. While the project is able to take advantage of lower cost gas prices in the second half of project life, discounting weights the immediate benefits of lower oil prices far more heavily than it does lower cost gas supplies in the out years. For the same reasons, the gas option produces the lowest NPV.

Once uncertainty with respect to future fuel prices is introduced into the analysis, the results change fundamentally. Assume that oil prices next year could be $1.18/ mmbtu, with a 30 per cent probability, $0.79, with a 40 per cent probability, or $0.39, with a probability of 30 per cent. Next year's weighted average oil price is thus $0.79/mmbtu, and as before is expected to increase at a rate of 5 per cent per annum. Note that the expected value of future oil prices is identical to the prices assumed in the certainty case. The key implication, therefore, is that with prices distributed symmetrically around the mean price, the project's calculated NPV is independent of oil-price volatility.

Table 6.4   Base case economics: oil, gas and dual fuel (in $m)

A. Cash flows under certainty

	Years										
	0	1	2	3	4	5	6	7	8	9	10
R		1.16	1.21	1.27	1.34	1.40	1.47	1.55	1.63	1.71	1.79
C_G	−2.5	1.00	1.00	1.00	1.00	1.00	1.00	1.00	1.00	1.00	1.00
C_O	−2.5	0.79	0.83	0.87	0.91	0.96	1.01	1.06	1.11	1.16	1.22
C_D	−3.0	0.79	0.83	0.87	0.91	0.96	1.00	1.00	1.00	1.00	1.00
NCF_G	−2.5	0.16	0.21	0.27	0.34	0.40	0.47	0.55	0.63	0.71	0.79
NCF_O	−2.5	0.37	0.39	0.41	0.43	0.45	0.47	0.49	0.52	0.54	0.57
NCF_D	−3.0	0.37	0.39	0.41	0.43	0.45	0.47	0.55	0.63	0.71	0.79

B. Cash flows under oil price uncertainty

Fixed oil plant

	0	1	2	3	4	5	6	7	8	9	10
R_O		1.16	1.21	1.27	1.34	1.40	1.47	1.55	1.63	1.71	1.79
P_{OH}	2.50	1.18	1.24	1.30	1.37	1.44	1.51	1.58	1.66	1.75	1.83
P_{ON}	2.50	0.79	0.83	0.87	0.91	0.96	1.01	1.06	1.11	1.16	1.22
P_{OL}	2.50	0.39	0.41	0.43	0.46	0.48	0.50	0.53	0.55	0.58	0.61
P_{AVG}	*2.50*	*0.79*	*0.83*	*0.87*	*0.91*	*0.96*	*1.01*	*1.06*	*1.11*	*1.16*	*1.22*
NCF_H	−2.5	−0.03	−0.03	−0.03	−0.03	−0.03	−0.03	−0.04	−0.04	−0.04	−0.04
NCF_N	−2.5	0.37	0.39	0.41	0.43	0.45	0.47	0.49	0.52	0.54	0.57
NCF_L	−2.5	0.76	0.80	0.84	0.88	0.93	0.97	1.02	1.07	1.12	1.18
NCF_{AVG}	*−2.5*	*0.37*	*0.39*	*0.41*	*0.43*	*0.45*	*0.47*	*0.49*	*0.52*	*0.54*	*0.57*

Flexible project

	0	1	2	3	4	5	6	7	8	9	10
R_F		1.16	1.21	1.27	1.34	1.40	1.47	1.55	1.63	1.71	1.79
C_H	3.0	1.18	1.00	1.00	1.00	1.00	1.00	1.00	1.00	1.00	1.00
C_N	3.0	0.79	0.83	0.87	0.91	0.96	1.00	1.00	1.00	1.00	1.00
C_L	3.0	0.39	0.41	0.43	0.46	0.48	0.50	0.53	0.55	0.58	0.61
C_{AVG}	*3.0*	*0.79*	*0.75*	*0.77*	*0.79*	*0.81*	*0.83*	*0.84*	*0.85*	*0.86*	*0.87*
NCF_H	−3.0	−0.03	0.21	0.27	0.34	0.40	0.47	0.55	0.63	0.71	0.79
NCF_N	−3.0	0.37	0.39	0.41	0.43	0.45	0.47	0.55	0.63	0.71	0.79
NCF_L	−3.0	0.76	0.80	0.84	0.88	0.93	0.97	1.02	1.07	1.12	1.18
NCF_{AVG}	*−3.0*	*0.37*	*0.47*	*0.51*	*0.55*	*0.59*	*0.64*	*0.71*	*0.77*	*0.85*	*0.92*

Source: Kulatilaka and Marcus (1992).

Key: R = revenues; C = costs; NCF = net cash flow; H = high; N = normal; L = low; AVG = average.

The impact that uncertainty has on the higher cost dual-fuel option is far more interesting. Expected energy costs averaging out across the three investment options are far lower, and the corresponding cash flows higher, in all periods. The NPV of the flexible option increases by nearly $0.7 million, to $0.64 million. Contrary to the original analysis, the dual-fuel option now dominates both the oil and gas technologies. Kulatilaka and Marcus conclude that 'the flexible technology produces a valuable symmetry: when oil prices follow the low path, the cost savings accrue fully to the firm. By contrast, when oil prices follow the high path, increases in oil costs are limited by the ability to switch fuels. In contrast to the cash flows for the fixed boiler option, the expected cash flow is actually a function of oil price volatility, even when that volatility is symmetric around the mean price. Higher volatility increases average net cash flow since benefits accrue when oil prices fall, but losses are limited by the price of gas.'

The same argument can be recast within a options-theoretic framework, the point of the original analysis. The dual-fuel investment in effect concedes to the electric utility a call option: the company can always purchase fuel inputs at the lower of the fixed gas cost or the price of oil. The holder of the option in effect has a right to acquire an asset for the lower of the strike price, here corresponding to the fixed price of gas, or the market price of the asset, the price of oil. In comparison with the oil-only investment, the flexible technology option results in cash flows that are higher in each period by the amount $\max(p_{oil} - p_{gas}, 0)$, that is, the price differential if positive or zero otherwise.

The point of the example is to show the utility was correct to pay a premium for the flexible fuel option since it permitted the company to always be able to source the lower cost fuel input. Flexibility is valuable and investors accordingly ought to be willing to pay for it. In one significant respect the example is too facile: at the margin, oil and natural gas prices would be expected to equalise once proper account is taken of differences in thermal efficiencies or the fact that oil entails storage costs while natural gas does not. This principle, known as thermal equivalence, would appear to vitiate the point of the example. There really is no need for investors to incur the higher capital costs connected with constructing the dual-fuel facility; the same degree of flexibility could have been engineered into the investment via the fuel purchase agreement. More to the point, to the extent the utility company has the ability to source a single fuel, say, natural gas, at the lower of the market or thermal equivalent price, then it effectively reverses Kulatilaka and Marcus' conclusion. The natural gas option now makes the best economic sense. Investors will have saved on the higher up front capital costs, while minimising fuel costs period by period. The utility has, in effect, acquired for the duration of the fuel contract an option always enabling it to purchase natural gas at the lowest cost.

This description is far more representative of current market conditions than the example devised by Kulatilaka and Marcus. Upstream natural gas producers are prepared to enter into long-term contracts that guarantee buyers will be able to source their fuel supplies at the lower of gas or thermal equivalent prices. Producer interests derive from having secure long-term outlets for their production, in the case of natural gas the *sine qua non* for obtaining finance to develop the project. Nor is the willingness of the gas markets to enter into such commitments reflective of unique market developments that might eventually change, leaving producers with a substantial, permanent portfolio of low price sales contracts while guaranteeing purchasers ongoing windfall cost savings. Indeed, if this were likely to be the case, it could be described as reverse take-or-pay, with purchasers and producers having swapped places.[9]

So long as market conditions remain the key determinant of fuel prices, it is unlikely that price discrepancies could persist for very long. In which case, flexibility in the sense used by Kulatilaka and Marcus needs to be redefined; in the case to hand, it means being able to secure least cost fuel supplies via contractual pricing arrangements. Seen this way, investors will have been able to procure option-like benefits but without having to incur the additional costs such flexibility theoretically entails. It should be equally clear that traditional cash flow methods are perfectly adequate for assessing the economic benefit of investments underpinned by such contractual flexibility.

TRADITIONAL PROJECT FINANCING: MIDDLE EASTERN LNG

The present case study assesses the economic rationale for major LNG projects in the Middle East. The initial justification for a project in Qatar, now the region's pre-eminent producer and exporter of LNG, was twofold: to develop a vast natural gas resource through liquefaction, in order to supply the east Asian market, the fastest growing in terms of both current and prospective LNG consumption. In fact, east Asia was the sole logical export outlet, notwithstanding interest shown by Italy and other south European countries in the early stages of the project (Shell, 1992). So long as the European market remained subject to tight regulatory control, there was always some scope for including a high cost Middle Eastern LNG tranche so as to diversify regional gas supply. Once it became clear, however, that the European gas market would be liberalised the possibility of importing expensive supplies from the Middle East disappeared from southern Europe's gas agenda.[10]

The power sector is and remains the principal end-use market for LNG. Prior to the recent financial crisis, east Asian electricity demand was expected to continue to grow rapidly. Regional power demand, on the other hand, could be met just as easily using lower cost indigenous fuels. Both China and India, for example, are major coal producers, so there was no necessary reason why the bulk of future Asian power demand could not have been covered by traditional fossil fuels.[11] There were, moreover, numerous potential competing projects, either within the Middle East (Oman, Egypt, and Yemen, for example) or the east Asian region itself (Australia, Brunei or Indonesia, all of which are major producers and exporters of LNG).

As long as economic considerations dominated regional fuel source decisions, coal should remain the fuel of choice. Higher income east Asian countries, however, appear to favour natural gas mainly for environmental reasons, although whether the commitment to stringent environmental targets will survive current financial pressures is an open question. At minimum, the crisis should result in a significant reduction in prospective regional gas demand, a concurrent decline in LNG imports, increased competition among existing LNG exporters with a potentially significant negative impact on prices and, by extension, project cash flows.

By way of background, we might note that world LNG trade amounted to 74.2 million tonnes per years in 1996, roughly one-quarter of total global gas trade (British Petroleum, 1997). East Asia is the major market for LNG: regional imports account for more than three-quarters of world LNG trade with Japanese imports alone representing just under two-thirds of the global total. On the supply side, Indonesia is far and away the world's largest producer and exporter of LNG, accounting for 35 per cent of the world total. Algeria is the next largest exporter, with a 20 per cent market share, while Australia and Brunei each account for about 10 per cent of global LNG shipments. Abu Dhabi was until recently the sole Middle Eastern LNG producer and exporter. World liquefaction capacity currently amounts to 92.6 million tonnes per year, with an additional 23 million tonnes per year currently under construction. Planned and speculative liquefaction capacity is estimated at a further 104.8 million tonnes per year (Dols and Page, 1998). Many of these latter projects were conceived against the expectation of continued strong economic and energy growth in east Asia, coupled with the emergence of countries such as India, Thailand, and Turkey as significant potential future purchasers. How many of these investments will in fact materialise is an open question.

The Far East is, as already noted, the sole logical destination for Middle Eastern LNG production, notwithstanding the initial interest shown by some European countries. Two export outlets for project output would, of course, have materially enhanced the attractiveness of the project. An analysis carried out by the International Energy Agency (1991) shows why such two-way trade was never on the cards, unless energy security dominated price considerations. According to the IEA, the delivered cost of LNG from the Middle East to western Europe was, at $4/mmbtu, the highest of any of the options surveyed. The cheapest incremental sources of supply were, naturally, those closest to hand: marginal expansion of existing Russian or Dutch supplies would have cost only $0.25–0.50/mmbtu.

The lowest cost LNG supplies identified in the IEA study originate in Algeria: removing bottlenecks in existing facilities would result in a delivered cost of $0.50–1.00/mmbtu. Greenfield LNG ventures geared to the European market include new projects in Algeria and Nigeria. Given the political and social difficulties being experienced in both countries it is hard to imagine either would qualify in current circumstances. In short, there were numerous cost effective options to construction of a new LNG plant in the Middle East.

The Asian LNG market is dominated by three countries: Japan, South Korea and Taiwan. Projections compiled just prior to the financial crisis envisaged continued rapid consumption growth in each of these countries; post-2000 the market was expected to expand to include India, Thailand and China. In keeping with these forecasts, the Asian Development Bank predicted the total financial requirements connected with continued growth of regional LNG consumption at $70 billion to the end of 2010 (Symon, 1997). The projections given in Table 6.5 are broadly representative of the expected pre-crisis evolution of regional gas demand.

Even abstracting from the negative impact the crisis is bound to have on regional import requirements, the inclusion of India and China looks speculative. Both are major coal producing countries, with coal accordingly accounting for the bulk of local electricity generation. Moreover, currently planned power projects are based mainly upon coal, as might be expected

Table 6.5   LNG demand in Asia Pacific (million tonnes)

Year	Japan	S. Korea	Taiwan	India	Thailand	China	Total
Actual							
1994	41.2	5.7	2.2				49.1
1995	41.5	6.8	2.5				50.9
1996	46.6	9.5	2.5				58.5
Forecast							
2000	54–55.5	13.8–17.5	6–8				73.8–81
2005	57–60	15.5–23	9–12	2.5–5	2–6	2–4	86–110
2010	60–70	18–28	12–14	2.5–5	2–6	2–4	94.5–127

given cost and local industrial development considerations.[12] Finally, neither India nor China have the same credit standing as traditional LNG importers. Sponsors accordingly appear unwilling to commit to project development based solely upon long-term contracts entered into with utilities in non-traditional importing countries. Over the longer term such countries could represent expansion potential for existing projects.

Finally, there is the important question of what impact the current financial crisis is likely to have on current and future LNG import requirements. We noted in Chapter 1 that Korean utilities have started to scale back imports and on current trends are expected to eliminate spot deliveries entirely this year. Contract purchases are, of course, higher cost than are spot supplies. Since energy exports are typically invoiced in US dollars, the recent 35 per cent depreciation of the Korean currency is adding further to the country's total energy bill. The extent to which Korean energy policies threaten future economic well-being depends primarily upon the pricing formulas contained in existing import contracts. Earlier agreements were underpinned by guaranteed minimum real purchase prices. More recently, in exchange for a commitment to double project offtake, Ras Laffan agreed to drop the guaranteed floor price that was part of the original sales and purchase agreement (Gotaas-Larsen, 1997).[13] The new formula links import costs to competing fuel prices, namely, Japanese crude costs.[14] In which case, the planned elimination of spot in favour of contract supplies is entirely cosmetic: Korea Gas Corporation is in effect substituting contract for spot supplies while paying spot for contract supplies.

The ultimate impact on the Korean economy now appears to be more manageable when compared to the alternative of having to swap higher for lower cost supplies coincident with depressed competing fuel prices, as was the case over most of 1997, and a permanently lower value for the won, *vis-à-vis* the dollar. The potential detrimental impact on project economics, and by extension, debt capacity, has not gone unnoticed. Indeed, recent developments in the LNG market bear comparison with the financial difficulties that confronted creditors in the mid-1980s, when owing to a fundamental change in market conditions purchasers either repudiated or renegotiated high cost coal import contracts. The quintessential example of this process was that Japanese utility companies completely rewrote high cost contracts with Canadian coal producers following the collapse of oil prices in the mid-1980s, notwithstanding these utilities had an equity stake in the same projects. It might be noted in this connection that Japanese companies hold similar equity interests in recent LNG projects. For example, Marubeni and Mitsui each have a small stake in Qatargas as do Nisho Iwai and Itochu in Ras Laffan. It has been suggested, notwithstanding the experience of the coal market in the 1980s, that recent and prospective commercial changes argue strongly in favour of enlarging the investor base of future LNG projects as a way of mitigating such risks (Dols and Page, 1998).

All of the available evidence indicates that Middle Eastern projects are well placed to supply the Far Eastern market. Among large grassroots projects, Qatar is one of the lowest cost potential suppliers. Only Borneo has a lower delivered cost, with most other potential projects, including, significantly, those within the east Asian region, typically above the estimated cost of Middle Eastern exports. Only two projects showed lower delivered costs to Japan and both not surprisingly are expansions of existing operations (Poten and Partners, 1991). For major grassroots projects capable of supplying the Far Eastern market, the variance in capital cost estimates is narrow, although two regional projects, above all, Natuna (Indonesia), appeared to enjoy a clear cost advantage over Qatar. In contrast, production and liquefaction costs tend to be lower in the Middle East, while shipping costs, owing to the longer distances, tend to be twice as high. More positively, production and supply costs are lower for Gulf projects, although Indonesian projects would appear to enjoy significantly lower liquefaction costs compared with a Middle Eastern project.

While Middle Eastern project economics are mixed with respect to a comparably sized Indonesian project, there are several other factors that tip the scales in favour of the Gulf. First of all, resource availabilities are far higher in, say, Qatar, that any other potential LNG site. Estimated gas reserves amount to 150 trillion cubic feet, more than 2.5 times the high estimate of Natuna's gas reserves and nearly 20 times Kalimantan. And second, Qatar stands out in terms of the importance of gas co-products, above all, condensates, 22 per cent versus nil for Natuna. Under any circumstances these liquids provide extremely valuable production revenues and go a long way towards explaining why upstream development costs in Qatar are significantly lower than elsewhere.[15]

Indeed, the gas stream is so rich in liquids it was possible to divide the upstream from the downstream phase of the project and finance each individually. In other words, the gas field was developed as a separate condensate project while the downstream was financed as an LNG export project.[16] Condensate credits, depending upon their valuation, could easily have exceeded production costs. Because of their higher API gravity, it is possible to obtain from condensates higher yields of more valuable refined products, gasoline, for example, hence the reasonable expectation they will command a premium to local crudes. It turns out, however, North Field condensates have a very high sulphur and mercaptan content; it is, of course possible to upgrade the condensates, but that involves substantial additional cost (Pollio and Hart, 1994). The key point for present purposes is that, regardless of their quality, liquids enhance the competitiveness of a Qatari project *vis-à-vis* other possible investment venues.

The relevant financial and economic data are summarised in Table 6.6 where, for analytical purposes, it is assumed that the principal competitor to a grassroots Gulf project is in Indonesia. In comparison with the east Asian

Table 6.6   Comparative LNG project economics: Qatar and Natuna, Indonesia

Characteristic	Qatar	Indonesia (Natuna)
Capital costs (M)	$6 380	$5 328
Annual operating costs (M)	$158	$312
Gas reserves (trillion cubic feet)	150+	40–60
Existing LNG projects	no	yes
Political risk	moderate	moderate/high
Sponsors	major oil companies	major oil companies
Market prospects	good	good
Financial enhancements	extremely favourable to sponsors	typical for similar projects
Bankable	yes	yes

alternative, the Qatari project looks reasonably attractive, although capital cost estimates are nearly $1 billion higher.[17] More positively, Gulf project operating costs are lower and gas reserves nearly three times higher; project economics are further enhanced by significant co-product availability. Political risks, even from the perspective of the mid-1990s, appear no worse or possibly even lower in the Gulf region, while energy diversification considerations would seem to favour a Gulf over an Indonesian project.

Finally, financial concessions are far more attractive in the Gulf project, providing investors with greater assurance that project hurdle rates will be met. More to the point, the government of Qatar agreed in effect to subsidise project feedstock costs if the project failed to meet a pre-determined investment hurdle rate, to provide additional financial support via a natural gas liquids sales convention that overrides the contractual processing fee with actual sales proceeds and, if required, by extending the original ten year tax holiday by an additional two years.[18] It should be noted that the magnitude of the financial concessions depended upon project scale. For example, if the original Qatargas project was increased, as it was, from two to three trains, some of the financial concessions were modified or withdrawn. Financial concessions made by the state to the project were to be treated as loans, subordinate to the project's commercial debt and repayable only if improved project performance permitted. In fact, the joint venture agreement stipulates a number of tests covering the provision of financial support and the method and amount of clawback. Finally, project economics were further underpinned by a minimum floor price conceded by the Japanese utility company importing the gas, and set at a level ensuring that project sponsors earned their required rate of return.[19] All in all, the project provided a number of extremely attractive technical and financial features for potential industrial investors.[20]

The ultimate merits of the project depend upon two other critical considerations. The first is that LNG design capacity, is always understated, owing to fairly conservative assumptions concerning the number of days over the course of the year the facility will be operational. Actual capacity utilisation normally exceeds design capacity, implying a proportional increase in project cash flow if the additional volumes are sold at contract prices. In the more likely event, such production will be disposed of at spot prices, with contract purchasers having a right of first refusal on such supplies. Contract purchasers also have what is known as a right of downward flexibility – the purchaser has an option to reduce contractual takes, although the timing and the volumes so affected are restricted.[21] Having exercised a right of downward flexibility, there is nothing to prevent the buyer from sourcing supplies at spot prices from the same or another project.

Spot sales are an important component of total gas imports – Korea, for example, sources roughly 44 per cent of its LNG imports on this basis (*Petroleum Review*, 1998) – since they have the practical effect of lowering weighted average gas acquisition costs. Downward flexibility provides purchasers with additional financial protection in that, during periods of low competing fuel prices, buyers can alter the supply mix with an attendant positive impact on the import unit value. The right of downward flexibility thus provides a cost-effective solution to the risk of repudiation, by conceding to contract buyers a mechanism whereby they can lower import costs within the framework of existing contractual arrangements. Such mechanisms, of course, can not address the problem of a permanent differential *vis-à-vis* spot purchases. Should the disparity persist for any length of time, the pressures to renegotiate are bound to prove overwhelming; indeed, this already appears to be happening.

Even a discussion as brief as the one just given highlights the significance recent developments are bound to have on the willingness and, accordingly, the terms that will apply to future LNG development. Three in particular stand out. The bulk of existing LNG projects are focused on east Asia. For projects geared to the Japanese market, the impact of the regional financial crisis on creditor interests is negligible. For projects focused on emerging Asian markets, the impact is of far greater concern; such fears were clearly behind the recent decision to place Ras Laffan bonds on a watch list. The second major source of concern is the loss of traditional commercial arrangements, minimum sales prices, for example, that ostensibly protect creditor interests in the event of a fundamental change in market conditions. Such mechanisms never made much sense, since in practice they are most likely to be repudiated precisely when their protective function is most urgently needed. And finally, if the forecasts given in Table 6.5 are to be believed, future industry growth depends upon project output being sold to purchasers of lower credit standing than historic buyers. In short, future LNG projects will be exposed to greater commodity and credit risks than financial markets have become accustomed to, with an attendant hardening of the terms of future credit commitments.

A FERTILISER PROJECT IN PAKISTAN

In contrast to Qatar, Pakistan is currently unable to borrow on its own guarantee, effectively shutting the country out of international financial markets. Local projects, regardless of their economic merits, will not qualify for financing unless they are backstopped by guarantees provided by creditworthy entities.[22] These would include, above all, international development agencies, the World Bank, the Asian Development Bank or the Commonwealth Development Corporation, for example. Any commercial loan tranche would of course have to be guaranteed by national export credit agencies (ECAs), the Export Credit Guarantee Department in the United Kingdom or the Export Import Bank in the United States. The direct coverage provided by ECAs plus the customary cross-default clauses common to World Bank loans effectively eliminate political risk from international loan transactions. In practice, ECAs allow lenders to reallocate the sovereign risk exposure to the country or countries providing the financial guarantees, thus averting any potential increase in bank capital.[23]

Financings backed by export credit agency guarantees are not without risk, however. ECAs can and have refused to compensate lenders for political risk events; indeed, ECAs have been known to demand repayment of claims previously disbursed. These are to be sure extremely rare contingencies, usually confined to those cases where loan authorisations were based upon inadequate, improper or illegal undertakings. A financing underpinned by a guarantee provided by a state agency which lacked the statutory authority to do so, what in law is called *ultra vires*, is a prime example. If lenders made the loan knowing, or having reason to believe, that the local guarantor was not authorised to furnish the guarantee, financial responsibility would ultimately rest with the lending syndicate. In the more usual case, lenders will have satisfied themselves as to the legal standing of the guarantor, with the *ultra vires* issue arising only subsequently. Lenders will normally adjudicate the issue; if they obtain a favourable verdict, ECAs will almost invariably honour the claim. There have, however, been instances when, even after having obtained a verdict favourable to the banks, the ECA refused to disburse the remaining balance or continued to demand repayment of previous disbursements.

The background to the case at hand is that the local project sponsor, an established and experienced producer of urea fertilisers with a 42 per cent share of the Pakistan market, is seeking financing to build an integrated manufacturing facility that will produce 445 000 tonnes/year (t/y) of di-ammonium phosphate (DAP) and 551 000 tonnes of granulated urea. The project is underpinned by strong local demand for both nitrogenous and phosphatic fertilisers, a portion of which was historically covered through imports. The project will eliminate the need for imports while generating a small exportable surplus. Both the DAP and urea production facilities will utilise established technologies appropriate for large-scale processes. Plant performance is underwritten by robust contracts. Finally, a 419 000 t/y second-hand ammonia plant, needed to feed both the DAP and urea plants, was purchased in the United States. For the main project sponsor, the new facility complements its existing operations; the joint venture will also make use of the company's established brand name and exploit its existing marketing and distribution networks. Phosphate rock, needed to produce phosphoric acid – a DAP production input – will be sourced in Jordan by a local producer, which has existing mine and DAP production facilities there. This company also has a 10.4 per cent equity stake in the venture.

An engineering, procurement and construction contract was signed with Klockner-Krebs, a Franco-German consortium. The facility is to be constructed under a fixed-price turnkey contract. The contract contains performance guarantees as well as a 12-month warranty which covers engineering, materials, spares and licences. Plant technology is standard for this type of operation. It will take roughly 32 months to complete construction of the plant. Project management as well as operational personnel are to be provided by the project's principal sponsor. The second hand ammonia plant, on the other hand, could be a source of some concern as it lacks any sort of performance warranties. Even so, many of the plant's key components have been refurbished or replaced, and many of these are covered by warranties of up to three years. Consultants retained to evaluate the facility conclude it is in satisfactory condition, better in fact than a plant recently acquired by a local competitor that is now up and running. Furthermore, the second-hand facility was known to have been operational up until fairly recently and was shut down and sold off owing to an explosion in its associated urea unit.

The development value of the project can be assessed across four main dimensions: job creation potential, balance of payments impact, local industrial development and impact on local agricultural productivity. The project is expected to create fewer than 1000 positions. Many of the new jobs will require recruitment of highly skilled workers; most, however, will be less technical and clerical, including additional staff to be recruited by the marketing and distribution departments. Net annual foreign exchange savings should amount to roughly $50 million per year once the plant is fully operational. The savings derive principally from the substitution of lower cost phosphoric acid for higher value DAP imports.[24] If the project is subsequently able to export some portion of project output, then the balance of payments impact will be proportionately greater. The project is located in Port Qasim in southern Pakistan, and will have the beneficial impact of promoting the growth and further development of the site as the country's leading industrial zone.

The promotion of local agricultural development is far and away the project's principal *raison d'être*. Fertiliser consumption in Pakistan, although having grown rapidly – by about 7 per cent per annum over the past two decades – is still low by international standards. For example, recent application rates in Pakistan are in the 90–100kg of nutrient per hectare range as against 340kg per hectare in the United Kingdom. Agricultural production will be boosted further by the increased proportion of phosphatic in total fertiliser application. The pricing of project output is more problematical. The government of Pakistan has in recent years phased out all direct subsidies, including those that formerly applied to fertilisers. Inconsistently, the government will continue to subsidise gas-based fertiliser production, the price differential being needed to ensure that the price of local supplies will not be undercut by imports. Only during periods of global economic weakness is there the risk that the cost of locally produced supplies will exceed the cost of imported urea.[25] To address this risk, the government is committed to guaranteeing a floor price of $250 for DAP, while the urea price will be set such that investors can expect to earn a minimum equity return of 20 per cent at 90 per cent capacity utilisation.

For present purposes we leave aside the issue of the relevant rate to be used to discount project cash flows. Instead, we shall consider some of the risks that could affect project economics. On the positive side, the project appears to enjoy strong sponsor support and will be managed by competent staff with considerable local experience in the production, marketing and distribution of fertilisers; the same expertise will be available in respect of training and the provision of technical and financial support. On the negative side, the principal risks relate to the supply of gas feedstock. A supply agreement was executed between the project and the Sui Southern Gas Company (SSGC) that provides for a total commitment of 75 000 mcf/d of natural gas, adequate for full capacity utilisation. SSGC, however, is prepared to offer firm sales covering only 40 000 mcf/d, with the balance to be supplied on an 'as and when available' (namely, interruptible) basis.

The firm portion of the supply contract limits the plant operating rate to 60 per cent of capacity, some 15 percentage points below the project's break-even threshold. The gas sales agreement stipulates that the volume of gas to be made available to the project shall always be set in keeping with 'allocations fixed by the government'. The key implication here is it is the government, not SSGC, that has determined the plant's firm allocation, in reality a distinction without a difference since the gas company is publicly owned. The Ministry of Petroleum and Natural Gas and SSGC claim that it is highly unlikely the project will ever obtain less than its full natural gas quota. Fertiliser plants have the highest priority in the allocation of gas supplies to the commercial sector. Moreover, future gas finds coupled with enhanced production from existing fields are said to favour areas served by the southern pipeline system, the network that supplies feedstock to the project.

Feedstock availability risks are compounded by cost factors. The price of natural gas is set by the government of Pakistan via a formula that links the cost of local supplies to the international price of fuel oil. Gas prices are thus subject to change in keeping with fluctuations in international oil prices. Gas used in the manufacture of fertilisers benefits, as noted above, from official subsidies, with the price fixed at around one-third the benchmark fuel oil price. Roughly 80 per cent of project feedgas supplies will be subsidised. Feedgas costs are to be set 6 months prior to startup, from which time prices will be fixed in nominal rupee terms over the next decade; the price includes transmission charges as well. The lack of a committed price introduces an element of uncertainty into the financial projections. On the not unreasonable assumption that prices applying to existing gas-based fertiliser operations will apply to the project to hand, these have been used in the base case analysis.

The final raw material input is phosphoric acid. The project entered into a long-term supply contract with a Jordanian company that will provide phosphoric supplies at a price equal to 97.5 per cent of the prevailing world price. The acid is to be shipped to the project plant site from the company's Jordanian facility. Limited production capacity raises the spectre of short-term shortages of phosphoric acid, in which case the facility will have to acquire needed supplies on the world market at international prices. Not only will the 2.5 per cent subsidy element have been lost, but, because these additional costs are incurred upfront, project economics are further penalised.[26] Nor, we might add, does the phosphoric acid supply contract contain any penalties in the event JPCM is unable to supply committed volumes.

The various issues just discussed highlight potential sensitivities. Table 6.7 presents five scenarios designed to assess the sensitivity of loan coverage ratios to different states of key project variables. Only those relating to gas supply and pricing would appear to require further comment. The first two scenarios focus on the availability and price of the project's feedgas supply. In the original gas supply agreement only 40 000 mcf/d of the total 75 000 mcf/d requirement were to be provided on a firm basis with the remainder supplied on an 'as and when available' basis.

Table 6.7   Key project risks and sensitivities

Gas quantity	*Scenario 1*: Assumes the project has access to only 77 per cent of its total (firm plus interruptible) gas allocation. At these levels, maximum utilisation is limited to 85 per cent of annual capacity.
Gas price	*Scenario 2*: Assumes the government does not adhere to its stated policy and the feedgas subsidy is withdrawn. Feedgas costs in other words are set equal to fuel gas costs; nor is the price fixed for 10 years as originally expected. In principle, higher gas costs should be offset by higher fertiliser costs.
Fertiliser prices	*Scenario 3*: Assumes that local and global fertiliser prices are 10 per cent lower than in the base case, while plant operating costs remain unchanged at base case levels.
Phosphoric acid prices	*Scenario 4*: Assumes that phosphoric acid prices are 10 per cent higher than in the base case, reflecting JPMC's inability to deliver under the contract forcing the project to source its phosphoric acid on world markets and prices.
Production levels	*Scenario 5*: based upon the operating history of FJFC's existing plant. Output is assumed equal to 100 per cent of capacity in the first year and 110 per cent thereafter. In the base case, production and capacity utilisation build more slowly, from 85 per cent in the first year of plant operation, to 95 per cent in the second year of project life and 100 per cent thereafter.

Source: CDC (1996).

Base case economics were predicated on the assumption that the probability of the project not receiving its full quota was extremely low. Thus Scenario 1 is a compromise: it assumes that even if the project fails to receive its full gas quota, supplies will still be adequate to ensure plant operation at 85 per cent of capacity, the project's break-even threshold.

This scenario while not completely unreasonable is, however, disingenuous. It actually begs the question of the importance of feedgas supply by ensuring that even under worst case assumptions the project will always have access to volumes sufficient to guarantee project viability. Recall the firm portion of the contract – the only gas volumes project sponsors and lending banks know will be legally available – provides for a capacity utilisation rate of 60 per cent, which we know is uneconomic. There are two ways this issue should have been handled. The first option would have been to run base case economics on contractually committed volumes only on the assumption that the project could never be certain of obtaining feedgas supplies greater than 53 per cent of the total requirement. This in effect corresponds to a worst case scenario; the first sensitivity accordingly should evaluate project economics based upon 100 per cent availability of feedgas requirements.

The second option accepts original base case economics but for sensitivity analysis purposes evaluates project economics solely in terms of contractual volumes. Either option seems more in keeping with the objectives of sensitivity analysis than the compromise ultimately adopted. On the other hand, the gas price sensitivity is exactly right, quantifying project economics in terms of the complete loss of official subsidies; the opportunity cost of gas is set equal to its price as fuel, as would be the case in mature natural gas markets.

The results of the sensitivity analysis are shown in Table 6.7. Project economics are most sensitive to the loss of subsidised natural gas prices, which has the effect of significantly increasing operating costs, and low output prices. In both scenarios, the coverage ratio declines from 130 per cent to 90 per cent of the outstanding loan value. Phosphoric acid and gas supply curtailments have a more or less equal but smaller impact on the coverage ratio: in both scenarios project cash flow still exceeds the loan value, with the former having a slightly larger impact than the latter variable. Not surprisingly, higher production volumes have a favourable impact on project economics, resulting in a more favourable coverage ratio than in the base case.

One question not directly considered in connection with the evaluation of this investment was the correct rate to be used to discount project cash flows. As we said at the beginning of this chapter, no distinction is drawn any longer between social and private discount rates; where considerations of economic efficiency are paramount, the latter provides the best estimate of the former. Yet, we know different procedures can and have been used to determine the 'correct' discount rate for an offshore project. Nor is this a matter solely of academic importance, for the discount rate ultimately chosen can have a significant bearing on whether the project is viewed as being economically viable or not.

The discussion in Chapter 2 suggested two variants of the basic model. The first, which could be called the international capital asset pricing model, recognises and values any diversification benefits that result from undertaking an offshore investment. The second, described as the total risk approach, ignores any value that might accrue to sponsors from making

Table 6.8   Results of sensitivity analysis

Scenario	Coverage ratio
Base case:	1.3
Scenario 1: Reduced gas supply	1.2
Scenario 2: Loss of feedgas subsidy	0.9
Scenario 3: Low output prices	0.9
Scenario 4: High phosphoric acid prices	1.1
Scenario 5: 110 per cent capacity utilisation	1.4

Source: CDC (1996).

an international investment. For implementation purposes, the former includes the observed correlation between the investor's home market and the target market, while the latter approach constrains the correlation to unity.

The Commonwealth Development Corporation (CDC) (1996) recently attempted to derive a theoretically correct discount rate for its international investments. The CDC's mandate allows it to acquire equity stakes in international projects as well as providing loans for project development. Derivation of an appropriate hurdle rate for the project just described appears to follow a total risk approach. Two financial concepts are evaluated, the first being the financial IRR, corresponding to the return appropriate to a commodity chemical project; the second is the equity hurdle rate, which takes account of the degree of project leverage. The basic inputs needed to derive the first financial concept follow directly from the CAPM. The risk-free rate of interest is assumed equal to 8 per cent, on top of which the CDC adds a country premium, 1.5 per cent in the case of Pakistan. The third critical input is the market premium, which for present purposes is defined as the equity premium multiplied by the sector beta or

$$\text{Pre-tax market premium} = [MP/(1 - \tau)]\beta$$

MP is the market premium, τ is the tax rate and β is the beta for the UK commodity chemical sector. MP is assumed equal to 8 per cent, while the tax rate and beta are given as 33 per cent and 0.86, respectively. On this basis, the required (financial) rate of return is equal to:

$$\text{FIRR} = 8.0 + 1.5 + (8.0/0.67)(0.86) = 19.8 \text{ per cent.}$$

The project would have to produce a 19.8 per cent rate of return if the project is to be economically justified; if we assume the ongoing rate of UK inflation is 3 per cent, then the real return corresponds to 16.3 per cent.

The next step is calculation of the required equity return, which unlike the FIRR takes explicit account of the degree of project leverage, namely, how much debt there is in the investment's capital structure. The first critical input is the after-tax unlevered ratio, namely, 19.8 per cent (1 − tax rate) or 13.3 per cent. To this figure is added the project's debt to equity ratio and the difference between the unlevered rate of return and the after-tax cost of debt. In the case to hand, the debt to equity ratio is 2.33 and the cost of debt is 11 per cent, the proposed project loan rate, which is equivalent to 13.8 per cent. Thus, the equity hurdle rate given actual project leverage is 27.1 per cent corresponding to a before-tax rate of return of 40.4 per cent. Note, finally, that these are theoretical hurdle rates, not the financial criteria actually used by the CDC to make international equity investments; the two can and do differ from each other.

Once we factor potential diversification benefits into the analysis, required rates of return change drastically. For example, if we assume the correlation between the UK and Pakistan stock markets is of the order of 0.35, the financial IRR declines to 13.1 per cent, while the required equity return falls to 22.2 per cent post-tax and 33.1 per cent pre-tax. Since within a CAPM framework only systematic risk is priced in the market, the lower the correlation between the home and target markets (that is, the greater the potential diversification benefits of an offshore investment) the lower will be required returns.

Investor practice, however, appears to favour evaluating offshore investments by disregarding any potential diversification benefits. To the extent that the 'correct' social discount rate corresponds to the one that would be used by the private sector, then it should be clear how large the implied financial penalty is compared with using the theoretically correct hurdle rate. In the project to hand, the difference is of the order of almost 700 basis points. The irony is that the financial parameters used by the CDC correspond more closely to those implied by the CAPM than the alternative formulation. This could mean one of two things. Either the CDC implicitly accepts that the CAPM provides the correct financial benchmark for its international investments or, alternatively, it uses the CAPM to subsidise international investments that could not be justified using the criteria applied by the private sector. Either way, one lesson is clear: emulating private sector financial practices to ensure that resources are efficiently allocated within emerging markets is easier said than done.

Notes

1   Debt instruments where the yield is linked to the behaviour of output prices are known as commodity bonds and are now relatively scarce in the energy sector (Priovolos and Duncan, 1991).
2   'A common starting point for negotiations on the procedure to be adopted in any particular credit agreement is that the lenders will provide the financial assumptions (that is, assumptions as to the discount rate and any relevant foreign exchange rates) and the assumptions as to product price but that the borrower will provide the more technical assumptions (e.g. as to capex and opex levels)' (Wood and Vintner, 1992).
3   It should be obvious that sponsors can, and do, set the production profile to create additional present value sometimes even to the detriment of ultimate recovery.
4   As noted above the principal departure from this practice occurred in the immediate aftermath of the second oil shock of 1979–80, and even then only briefly. By early 1983, lenders had reverted to pre-shock lending norms.
5   The risk-free rate appropriate for a long-term project investment is a long-term government security, say, a 30 year Treasury bond. For portfolio investors, the relevant rate would probably be a Treasury bill rate, corresponding to the shorter time horizon of the investment.

6   Recall that a project's internal rate of return corresponds to the discount rate at which the present value of future cash inflows equals cash outflows, that is, produces an NPV of zero. So long as the project IRR exceeds the investment's hurdle rate, NPV will be positive.

7   The principal shortcoming of Finnerty's model is the way the revenue function is modelled. This point is obvious if the model were used to assess the cash flow dynamics shown in Table 6.1. The production and, by extension, revenue profile of the oil project follows an inverted-V, reflecting a relatively rapid rate of resource depletion. Natural resource projects are unlikely to follow the revenue path described in this model.

8   The interest rate required by lenders to compensate for the higher risk of default can be determined by solving for the discount rate which equates the payoff at maturity with the current market value of the debt. The difference between the calculated return and the risk-free rate provides a direct measure of the required compensation. Weston and Copeland (1998) illustrate the relationship between leverage and interest premiums, while Brealey and Myers (1991) analyse the interactions between spreads, leverage and time to maturity. Required spreads, as might be expected, increase with both leverage and maturity. The functions, however, tend to be mirror images of each other. In other words, with maturities given spreads rise rapidly with higher leverage; with leverage given, the curves tend to flatten at longer maturities.

9   The take-or-pay problem referred to above followed deregulation, or liberalisation, of the natural gas market. Pipeline companies, the principal purchasers of gas in field markets, were expected to honour purchase commitments notwithstanding that contract prices were well above market clearing levels. Most of these agreements were eventually renegotiated to reflect the changed market conditions. The problem first surfaced in North America and more recently in the United Kingdom. With the European Union now having accepted that the same principle should apply across western Europe, these difficulties could easily be duplicated.

10 This result appears counter-intuitive since within an options-theoretic framework deregulation should be consistent with higher price volatility, meaning purchasers should impute a higher value to supply contracts. Given that the main end-use market for gas is the power sector, which has the widespread ability to generate electricity using alternative fossil fuels, cost not supply factors now dominate industry procurement strategy. 'The CIF cost of LNG delivered in Europe varies between $3.2–$4/mmbtu. These prices are acceptable in Asia ... but they are too high for Europe, which currently buys supplies at around $2.6/mmbtu' (Gas de France, 1997).

11 In 1996, China and India produced 1300.0 and 283.0 million tonnes of hard coal, respectively, and nominal amounts of soft coal. Coal consumption in the Asia Pacific region accounts for roughly 50 per cent of global coal demand; regional natural gas consumption, by contrast, represents just over 10 per cent of total gas demand. Asia Pacific gas demand is only around one-fifth of coal consumption. More positively, gas demand increased by 12.2 per cent between 1990 and 1996 versus 4 per cent for coal. The increase, of course, is measured from an extremely low base. Finally, we might note that the marker price for coal was only two-fifths the delivered cost of LNG to Japan (British Petroleum, 1997).

12 It might be worth noting that both China and India expressed interest in importing LNG from Qatar. The former, for example, in April 1994 signed a letter of intent with Ras Laffan to purchase an initial quantity of 2.5 million

tonnes per year with an option to double annual takes. Potential Indian imports, conceived within the framework of an Enron-sponsored power project, were discussed one year earlier (Gotaas-Larsen, 1997).

13   The implied increase in default risk was dealt with by the substitution of a Mobil guaranteed minimum floor price. More to the point, Mobil has agreed to provide to the project a revolving subordinated loan of up to $200 million at any time that can be used to pay the lesser of a debt service shortfall or the Kogas price difference. The former is the difference between the debt service payable on the due date and the amount of funds on deposit in a specially created debt service account. The latter is the difference between what Ras Laffan would have received but for the elimination of the minimum price and what LNG price the project would actually have received. The guarantee defines the minimum price as $1.90/mmbtu until 2009 and $1.65/mmbtu thereafter. The Mobil loan, which carries a 12 per cent interest rate, is subordinate to senior project debt but senior to equity distributions. See Rigby *et al.* (1997a) for a full discussion of the issue.

14   This feature now looks like becoming standard industry practice. There is, for example, no floor price in Shell's Oman sales and purchase agreement (Dols and Page, 1998).

15   'The key to the economics of long haul LNG is the value of the natural gas liquids (NGLs) extracted as co-products. A wet gas stream can yield $0.50–$1.00 per one thousand cubic feet in co-product credits; these co-product credits are the principal source of profit in most ventures (Stauffer, 1996). NGL value is further enhanced since they attract, as 'industrial' outputs, a lower rate of tax than does hydrocarbon production. Qatari NGLs consist mainly (two-thirds) of condensates with liquefied petroleum gases making up the difference. Official forecasts are for national condensate production to average 200 000 barrels per day in coming years. Condensate production is excluded from OPEC production quotas. Taking these forecasts at face value, Qatar's oil production (crude plus liquids) could in due course amount to close to one million barrels per day; this compares with the Emirate's current production quota of 378 000 barrels per day (*Financial Times*, 1998).

16   Qatargas upstream was financed with a 12 year, $570 million loan arranged by French banks. Loan proceeds were used to cover 70 per cent of the costs connected with drilling wells, building offshore platforms and piping vapourised gas onshore (Allen, 1996). The impact on potential lenders of the decision to finance the two phases of the same project separately is unclear. Generally speaking the decision appears to have narrowed lender interest owing to the loss of potential cross-subsidisation across the two project phases (Pollio, 1994).

17   It is worth noting that the capital cost estimates given in the table were developed in the early 1990s. Since then it has become clear that the investment costs shown for Natuna are far too conservative due, above all, to the project's remote location, water depth and the high carbon dioxide content of the gas stream (Thomas, 1998). The comparison is all the more apposite since Mobil has an interest in both projects. Mobil's stake in the Natuna project amounts to 35 per cent.

18   Some of these concessions are standard, while others are not. It is more significant that the concessions here are offered primarily to underwrite investor returns. In the more usual case, such concessions are made to secure project financing and it is customary that they are given by both the local parastatal and the foreign investor.

19 In fact, the floor price is escalated at the US rate of inflation resulting in a real price commitment. Contract buyers pay the higher of the market or the indexed floor price.

20 What the potential financial benefits to Qatar were is less obvious. The base case analysis assumed that with oil prices unchanged, the project would require the indefinite application of all the financial concessions agreed by the government. This means that the gas royalties and natural gas liquids sales revenues received by the government would always be well below their fair market value. Of course some part of these concessions would be recouped by the government via dividends paid by the project; the state through its national oil company held a 65 per cent interest in the project. True, the eventual enlargement of project scale meant that the foregone benefits would be reduced. Even so, base case economics provided little justification for the state undertaking the investment (Pollio, 1994). There are some potential offsetting factors, to be discussed below, suggesting that the base case understated potential project revenues. On the other hand, the financial crisis has drastically altered the medium-term outlook for east Asian gas demand and prices.

21 In the case of Ras Laffan, Kogas can reduce annual deliveries by 5 per cent per year for two consecutive years.

22 The Berne Union, the international ECA club, estimates that financing and guarantee facilities extended by members amounted $400 billion in 1995. Much of this amount was for large capital projects, like the two Qatari LNG projects and the one presently under discussion (Spence and Godier, 1996).

23 Capital adequacy is now determined in keeping with risk-weighted bank assets meaning that loans to countries with the same risk profile as Pakistan would attract a higher capital requirement. With, say, an ECGD guarantee, a loan to Pakistan would be classified as United Kingdom exposure, and reported that way on the lender's books.

24 The calculation excludes the economic losses associated existing feedstock subsidies and official minimum floor prices for both urea and DAP.

25 This is a rather doubtful proposition given that Pakistan's main competitors are gas-based fertiliser producers in the Middle East, above all, in the Gulf region. Gas production costs are considerably lower in the Gulf, and can be supplied at near zero prices while still covering local gathering and transmission charges.

26 In a present value sense, the additional costs are unlikely to be counterbalanced by the future availability of subsidised phosphoric acid.

7 Financing Infrastructure Projects

In the previous chapter we saw how basic financial principles can be applied to evaluate traditional project investments. For present purposes a traditional project is defined as one in which the debt used to finance the investment, and the revenue stream to repay it, are both denominated in the same currency. The project, in other words, lacks any foreign exchange exposure. A second distinguishing feature of traditional sectors is that lenders typically require project revenues to be accumulated in an offshore deposit account, with disbursements being under their direct control. Debt service generally has a priority claim, with the remaining amounts applied first to ongoing operating or maintenance expenditures. Once pre-set debt coverage tests are met, sponsors are free to distribute the residual as dividends.

Infrastructure investments normally, but by no means always, expose lenders to currency risk as most such projects are by design intended to meet local needs, the provision of power or transportation services, for example.[1] Airports are a classic example of projects that straddle the boundaries of traditional and non-traditional project sectors. Those that cater mainly or exclusively for international air travel will typically require carriers to pay landing and departure fees or other local expenses (fuel costs, for example) in US dollars. These receipts will be similarly escrowed, with creditors again being in a position to direct account disbursements. It is equally conceivable the project was initiated to cater for local air travel, and as might be expected the investment will create foreign exchange risk. In the more usual case, the airport will serve both domestic and international flights. Of critical importance to lenders is the proportion of total revenues generated by international enplanements, since these constitute the principal source of foreign exchange for debt repayment. The same broad considerations hold true, *mutatis mutandis*, for other cross-border infrastructure investments.

Otherwise, project lenders will typically require the borrowers to hedge the exposure or seek third party support in the event local authorities cannot provide the foreign exchange needed for debt service. The operative principle here is to have official third parties cover currency risks while leaving lenders to concentrate on technical, commercial and purely financial risks, precisely those they are best able to manage. For OECD projects with currency exposure – an independent power project in the United Kingdom financed, for example, in the US bond market – the principal risk is that an adverse change in the exchange rate will increase the local currency cost of meeting interest and principal repayments. If the project is vulnerable to such risks, it will almost always be possible to hedge the exposure via swaps or other derivative instruments and generally over the tenor of the debt. For similar projects in emerging market economies this is unlikely to be true, in which case, other risk management approaches will have to be adopted.

As a rule, infrastructure projects expose lenders to greater political and financial risks than do more traditional investments, such as oil or natural gas developments. On the other hand, Rigby (1997a) has argued that infrastructure investments, broadly speaking, entail lower technical risk than do projects in more traditional sectors. Technical risk in Rigby's model is measured across two key dimensions: project interactions and linkages. The first dimension concerns the connections that exist between or among project sub-systems. These can range from linear interactions – that is, where the sub-system interacts predictably only with preceding or following sub-systems – to complex interactions, 'which involve multiple reactions that create branching and feedback loops as sub-systems jump from one sequence to another'. Linear systems tend to be diffuse, supplies and raw materials are easily substituted, personnel specialisation is low, production steps are discrete and segregated, and failed components are easily identified and replaced. Complex interactions, by contrast, normally involve tight spacing of equipment, limited isolation of failed components, restricted substitution possibilities, unfamiliar or unintended feedback loops, a high degree of personnel specialisation across specific sub-systems and multiple and interactive controls.

The second dimension of multi-system risk addresses the interdependencies of different underlying sub-systems. Specifically, it assesses issues such as the degree to which sub-systems are time-dependent upon each other, the extent to which process sequences are invariant, the degree of flexibility or lack thereof in respect of being able to produce project outputs, the ability of the system as a whole to tolerate slack in supplies, equipment and personnel and, finally, whether redundancies and buffers exist.

Individual sectors can be sorted on the basis of how project sub-systems measure up against these criteria. As a general rule, project sectors involving tight linkages with low to moderate interactions are riskier than those where the reverse applies. On this basis, coal-fired power plants exhibit greater technical risk than do toll roads, but not as much as a refining project. Similarly, telecommunications projects owing to their relatively loose linkages are less risky than oil or gas pipeline projects. Such classifications are necessarily subjective, with the corollary that the results can not be pushed too far. Even so, they do conform, by and large, with *a priori* expectations concerning the nature of project risk.

Changing market pressures can fundamentally transform project structures and, accordingly, the nature of risks creditors have to face. Nowhere is this more true than in the electrical supply industry, where competitive factors virtually everywhere are altering basic project structures. One of the key attractions to both sponsors and lenders of early power investments was the existence of contractually binding capacity and energy payments that effectively underwrote project viability. Purchasers, moreover, were obligated to buy all power output tendered by the project. For power projects in emerging market economies sponsors derived the same, or even superior, benefits but at the cost of currency exposure. Such risks, however, were almost invariably mitigated through host government or third party guarantees.

It is ironic that the transformation of the global electrical supply industry has been driven in large part by the very success of independent power production. Owing to the progressive importance of competitive factors the future growth of generating capacity now depends more upon the provision of merchant than independent power supplies. This is strictly true in respect of OECD investments – several merchant power projects have already been financed in the United Kingdom[2] – and largely so elsewhere. We noted in Chapter 5, for example, that independent power production has largely or entirely eliminated supply constraints in many emerging economies. Not unexpectedly, this has provoked strong local pressures for host governments to revise downwards the tariff structures underpinning the initial investments. It seems reasonable to suppose that most such revisions will be handled less ineptly than the current Pakistani government has. Still, it would be remarkable if promoters were ever again to enjoy the extremely favourable terms governments were previously prepared to concede.

Merchant projects lack all the positive features common to independent power projects. Cash flows are exposed to what amounts to the same commodity (volume and price) risks as in more traditional natural resource projects. It may not be unreasonable to suppose, therefore, that in the future the more highly rated projects could be those with sponsors thoroughly familiar with merchant risks, oil producers or mining companies, for example.

Given these shifts, one interesting question is whether such pressures could result in a drying up of the project finance market, not unlike the reaction that set in following the oil price collapse of the mid-1980s. Whether it will is an open question. What we can say for sure is that the greater the exposure of project cash flow to market risks, the lower the loan value lenders will be prepared to concede to any individual investment. Equity, in other words, will have to account for more of total project cost than would otherwise have been the case. This fact alone might be expected to diminish investor interest in limited recourse debt to finance future infrastructure development. Such considerations, of course, relate uniquely to senior debt. Sponsors could, of course, achieve leverage targets by substituting subordinated for senior debt. This well help to maintain traditional debt–equity ratios, although at higher overall funding cost.

CURRENCY RISK MANAGEMENT

Exchange risk creates a mismatch between revenues and expenses that could impair the ability of the project to service its debts. Analysts differentiate transfer risk – the ability or willingness of host governments to permit their currencies to be converted into foreign currency – from exchange rate risk, which relates to fluctuations in the external value of the local currency. In the first instance transfer risk means that sponsors will not have access to hard currency needed to repay project debts or for distribution as dividends.[3] Transfer risk can also impair project viability via its impact on the cost of imported production inputs. To the extent that host governments refuse to permit conversion of local currency, the project may not be able to source raw materials essential for continued operation. Where transfer risk exists, most projects are unlikely to be rated any more highly than is the country in which the project is located. The second aspect of currency exposure arises from the fact that exchange rates are seldom stable, even when pegged to leading international currencies, the US dollar, for example. Owing to the long amortization periods characteristic of infrastructure projects, exchange volatility can adversely affect the project's credit risk profile. It can similarly impair project economics via its negative impact on project costs and ultimately cash flow.

Exchange rate dynamics have long occupied economic analysts, although it is fair to say no one dominant explanation has emerged. Indeed, there are a number of models of exchange rate determination, which we shall review briefly as a preface to our discussion of the ways currency risks are addressed in practice. Purchasing power parity (PPP), the first and perhaps best known theory of exchange rate determination, holds that prices of internationally traded goods must be equal at the margin. Otherwise, markets would arbitrage the resulting price discrepancies, buying the lower priced good and reselling it in the market where it will fetch a higher price. Exchange rate changes are the principal mechanism by which price equilibrium for internationally traded goods is achieved and maintained. A more sophisticated variant of the simple purchasing power theory asserts that exchange rates are driven not by prevailing price differences but rather in keeping with *expected* inflation differentials. In other words, the exchange rate of the country with the higher rate of inflation will depreciate by the expected difference in price performance.

Comparative price performance is typically proxied by national producer or retail price indices. Now, one of the principal difficulties connected with the use of national price data is comparability. If there are significant differences in the composition of market baskets as between goods and services, on the one hand, and traded and non-traded goods on the other, the use of national indices could result in rejection of PPP even if in reality it did hold. Several years ago the *Economist* magazine, as a way of eliminating the distortions caused by the use of national price data, introduced the concept of the Big Mac index. Since the Big Mac is the same everywhere, it ought to be possible to derive more consistent estimates of equilibrium exchange rates. When compared with market quotes, Big Mac implied parities would appear to provide a useful measure of the extent to which current exchange rates are over- or undervalued. Whether intended to be taken seriously or not, numerous empirical studies have attempted to test the validity of the Big Mac index. These results are worth considering not the least for the light they throw on the broader significance of purchasing power parity as a determinant of exchange rates changes.

Taken at face value, Burgernomics – currency valuations based upon international hamburger prices – suggests that any observed differences in Big Mac prices should be eliminated via offsetting exchange rate adjustments that result in what might be referred to as patty price parity (ppp).[4] Ong (1997), for example, recently tested this proposition over the period 1986–94 and found that the currencies of most emerging market economies in Asia and Eastern Europe were undervalued *vis-à-vis* the US dollar while the currencies of the higher income east Asian economies were uniformly overvalued. If correct, then the recent Asian financial crisis can be explained largely or entirely as a necessary correction to a fundamental misalignment of regional currency values. A second interesting finding is that over Ong's sample period the dollar was, with the exception of 1988 and 1994, consistently undervalued.

It is, of course, difficult to take these results at face value. In the first instance, PPP requires the absence of international trade barriers as a necessary condition for arbitrage of goods. Yet we know, for example, the Japanese government restricts imports of beef by, among other means, imposition of a sizable 50 per cent tariff. Since beef accounts for four-fifths of the commodity cost of a Big Mac, the extent to which official policies can distort the prices of internationally traded goods should be clear. Also, the Big Mac index makes no allowance for differences in tax rates that might similarly distort international comparisons. And finally, there is the possibility that productivity differences between countries could help to explain persistent departures of exchange rates from implied purchasing power parity levels.

The issue here is whether productivity growth increases more rapidly in the tradable goods sector than in the non-traded sector. If so, real exchange rates calculated on the basis of price indices that include both traded and non-traded goods will cause the currency of the country experiencing relatively faster productivity growth in the former sector to be systematically overvalued relative to its PPP level. There is, accordingly, no reason why PPP should hold once non-traded goods are factored into the analysis: except in very special circumstances such goods can not be expected to command the same price in different countries.

Now this is precisely the main shortcoming of the Big Mac index: with the traded component accounting for no more than 6 per cent of its retail price, the Big Mac is by definition a non-traded good and accordingly is not subject to the law of one price. On the other hand, if it were possible to exclude all non-traded costs from the comparative price, it should be possible to eliminate the productivity bias that may be obscuring the underlying relationship. To this end, Ong (1997) constructs what he calls the no frills index (NFI) and demonstrates that once allowance is made for productivity differentials, the adjusted BMI is a better lead indicator of actual exchange rates. The differences between the BMI and NFI are quite large, especially among the emerging market economies included in Ong's sample. In the case of China, for example, the BMI implies an exchange rate against the US dollar 80 per cent above the actual rate; applying the no frills index reduces the error to only 6 per cent. For the OECD economies, the margin of error, although still present, is much smaller.

We noted in Chapter 1 that the Fisher effect is widely used in financial analysis to explain the link between nominal and real interest rates. Foreign exchange analysts have used a variant of this relationship, known appropriately enough as the international Fisher effect, to explain exchange rate changes. The argument here is that currency appreciation or depreciation can be explained primarily in terms of changes in nominal interest rates. The theory predicts that currencies issued by countries having relatively low nominal interest rates ought to appreciate while countries having high nominal interest rates should experience depreciation. Again the key underlying assumption is that there are no restrictions on the free movement of capital so that arbitrageurs could exploit any net profit opportunities. On this basis, the law of one price requires that international interest rate differentials on securities of equivalent risk and maturity should uniquely reflect inflation differentials.

Yet a third line of inquiry leads to the conclusion that differences in interest rates between countries give rise to forward market discounts or premiums. According to the theory of interest parity, by which name this approach is known, any interest rate differences applying to equivalent securities will generate forward premiums or discounts of equal but offsetting magnitude.[5] In other words, forward rates adjust by an amount just sufficient to eliminate any interest differentials that might exist between two markets. If, by contrast, net profit opportunities existed, arbitrage ensures that parity would eventually be restored. Several commentators also regard forward rates as predictors of future spot rates. If correct, then future spot rates could be predicted either by focusing on the prevailing forward rate or, equivalently, the underlying interest differentials.[6] Of course, not all analysts are convinced the expectations hypothesis is correct, with the corollary that it will not be possible to use the information contained in the relevant variables to forecast future exchange rates.

Whatever the truth, the simple fact is that it is extremely difficult to predict foreign exchange rates, meaning that lenders will invariably require this particular source of risk to be eliminated from the financing. There are several ways in which this can be accomplished. Perhaps the simplest would be for sovereign governments or international development agencies to underwrite exchange risk via the provision of foreign currency payment guarantees. Of course, the potential scale of such guarantees could be so large as to threaten existing cash reserves, credit standing or financial stability of the guarantor agencies. Some would of course see in this the origins of the east Asian financial crisis, with the magnitude of official commitments having undermined national credibility, which in turn triggered the crisis.[7]

A second obvious source of mitigation is the use of derivative instruments, swaps, futures, options and so forth. From a creditor's standpoint the crucial issues here are the particular hedge mechanism used and the creditworthiness of the counterparties to the transaction. While acknowledging their potential advantages, financial analysts are surprisingly ambivalent concerning their actual value. Rigby (1997), for example, concedes the potential risk management benefits connected with the use of financial derivatives, but goes on to assert that as a general matter their value in project finance debt is restricted by currency coverage, contract length limitations and cost. Blaiklock (1998) is even more deprecatory: 'infrastructure projects are invariably complex without introducing such tools. Further, while the funding of a project with loans might embrace some flexibility to meet unexpected events, capital market instruments are relatively inflexible. To change the terms and timing of a swap might be very expensive to undertake if needs be. Contractually such swaps are related to a market and not the project in question, and great care therefore should be exercised before including them in a project financing structure.'

In many instances exchange rate risks have been successfully shifted to third parties, for example, purchasers of project output, raw material suppliers or government agencies. In each of these cases the third party essentially agrees to insulate project economics from currency risk through indexation of receipts or payments. The practical result is to shift the credit risk from the project to the guarantor, whose own credit standing may thus be impaired. Of course, the presumption here is that guarantors have 'deeper pockets' than does the project; otherwise, the net result would be to replace a weak credit with an even weaker commitment. Finally, we might note that where they exist borrowers can pledge offshore receivables to mitigate currency exposure; where they do not it may be possible to engineer substitute foreign currency payment mechanisms with the same result. By mitigating a key project risk, a project financing secured by hard currency receivables can also expect to lower overall funding costs (Pollio, 1993).

Substitute payment mechanisms, where they can be created – admittedly no easy task – provide the same degree of protection.

Rigby (1997) identifies five key criteria against which the mitigation of currency risk should be measured, three of which relate to project structure – and thus under more or less direct sponsor or creditor control – while the remainder relate to external factors, for example, the content of domestic economic and financial policy in the country in which the project is located. Since the latter are country-specific, each will have to be evaluated on its own terms. With respect to project-specific issues, one of the most critical is the extent to which the project's revenue stream is indexed to inflation or fluctuations in foreign exchange rates. Indexation, of course, is not without risk to project economics. The resulting automatic increase in output prices, potentially quite large, can have significant direct and indirect effects. In the first instance, it could reduce prospective consumption growth rates to the detriment of future project cash flow; equally important, it could trigger a political backlash making it difficult if not impossible to implement subsequent rate increases, even if contractually mandated.

A second equally significant source of creditor protection derives from restrictions on the distribution of dividends to project shareholders. In most loan agreements, cash distributions are permitted only after pre-determined debt coverage ratios have been met or surpassed.[8] Yet another way creditors can mitigate currency risks is through the creation of a debt service reserve. The magnitude of the reserve will depend upon past exchange rate performance, such that the larger past currency devaluation have been, the larger will the reserve required to address such risks have to be. As a general matter, exchange risk and domestic economic and financial stability go hand in hand; stable domestic policies are consistent with lower currency risk. On the other hand, exchanges rates can and have depreciated by amounts well beyond market expectations. One is hard pressed, for example, to cite obvious policy failures among leading east Asian economies prior to the eruption of the regional financial crisis in mid-1997; until then most of these same countries were widely regarded as paragons of economic virtue.

RISK CHARACTERISTICS OF INFRASTRUCTURE PROJECTS

The quantification and mitigation of project risks serves the very same purpose in the evaluation of infrastructure as it does in more traditional project sectors, namely, to address any risks that might impair the certainty of project cash flow. Non-recourse infrastructure debt is thus no different from loans to more traditional project finance sectors. In both instances, lenders' sole security in the financing lies in project assets and the cash flow they are meant to generate. The relevant risk factors for leading infrastructure sectors are given in Table 7.1. For comparative purposes, the risk factors appropriate to more traditional project finance sectors are also shown, in the case to hand, mining investments; these are the same factors discussed at length in previous chapters. The entries given in the table are those used to establish ratings for individual project bonds; the very same factors would apply in the case of limited recourse loans. It goes without saying that these criteria need to be supplemented by an evaluation of the characteristics specific to individual project investments. Only in this way will it be possible to arrive at a balanced assessment of default risk.

For traditional project finance sectors, the key risks relate first and foremost to the commodity itself. Limited recourse markets generally prefer precious to industrial metal projects unless the latter contain valuable co-products (usually precious metals) that enhance prospective cash flow. A second key risk is the quantity and quality of ore reserves, and whether the project presents any special construction or operational problems in either its mining or milling stages. As always, mining and processing technology must be standard otherwise lenders would refuse to accept the associated risks. Finally, there is the issue of the extent to which the project is exposed to political risk. As a general matter lenders are unwilling to accept such risks, so it is extremely unlikely a given mine project would qualify for funding if such risks had not been properly addressed.

Table 7.1 Principal project risks by industrial sector

Mining	Power		Pipelines
	Independent	Merchant	
Commodity	Power purchase agreement	Production inputs	Transportation – throughput contracts
Construction	Competitive production costs	Competition from new generators	Market served
Reserve	Fuel supply arrangements	Electricity substitutes	Competitive position
Operating plan	Technology	Legal/regulatory environment	Commodity
Financial/legal structure	Purchaser	Plant productivity and competitiveness	Regulatory
Country	Coverage ratio		Operating – technology
			Construction

Source: Standard & Poor (1997)

Country risk issues, by and large, have to be settled at the outset if limited recourse markets are to consider funding pending infrastructure projects; hence the absence of any explicit reference to political risk. On the other hand, contracts – whether purchase agreements in the case of independent power projects or throughput agreements in respect of pipeline investments – occupy pride of place. Competition emerges as a second risk factor in all infrastructure sectors, being rated slightly more highly for power projects than for pipeline investments. Note, finally, that competitive issues apply equally to independent and merchant power projects, although the natures of each are fundamentally different in the two market segments.

POWER PROJECTS

Table 7.2 identifies the key features of five recent US power projects which illustrate the principal risks faced by creditors when funding such investments. Three of the five projects are co-generation plants; one is in California, the other four are located on the US east coast. Two of the projects are owned by the same company. Three projects are gas-based while the other two use coal as the fuel input. The projects vary in size from 46–330 MW, while base case debt coverage ratios are uniformly 1.4 × or higher, but can be as low as 1.2 × for the low price scenario. For present purposes, interest lies primarily in the structure of the power purchase agreement – which sets the revenue line – and the fuel purchase agreement, which affects operating margins and by extension the scale of project cash flow. Before reviewing these variables, it seems worthwhile to summarise the technical characteristics of the individual projects, not the least because two rely upon new and largely unproven technologies (Panel C).

For projects that rely upon established technologies technical risk, not surprisingly, is rated as being low. Rosemary is the sole exception, although strictly speaking it is operational, not technical, risks that are the principal source of concern. Since it commenced commercial operation in 1990, the Rosemary plant has been plagued by numerous forced outages. True, the plant has yet to breach the 25 day outage allowance stipulated in the power purchase agreement. Nor it might be noted are the outages in any way connected with the plant's dual-fuel capability, which allows the facility to generate power using either natural gas or fuel oil with the additional capability of being able to switch fuels while on-line. The Rosemary plant is also inefficient, having a heat rate well below that of other plants in the sample. This constitutes a second potential threat to the project's revenue stream.

The assessment of the operating performance of the two plants that depend upon new generating technologies or equipment is far more encouraging. Fluidised bed technology, which underpins the Colver plant,

Table 7.2 Comparison of five US power projects

	Carson Ice-Gen	Colver	Indiantown	Rosemary	Brandywine
Owner	Central Valley Financing Authority	Inter-Power/Ahlcon Partners		Panda Interfunding Corporation	Panda Interfunding Corporation
Location	California	Pennsylvania	Florida	North Carolina	Maryland
Startup date		1995	1996	1990	1996
Description	46 MW combustion turbine and a 53 MW CC plant	102 MW bituminous waste, pyroflow fluidised bed boiler	330 MW pulverised coal, cogen facility	180 MW CC, natural gas, cogen facility	230 MW CC, natural gas, cogen facility
Annual capacity factor (%)	95.00	91.60	97.50	90.00	
Total energy delivered/produced (MWh)	479 494	777 300	2 553 155	90 629	
Steam sales (mmbtu)	14 480		525 000[2]	351 674[2]	
Fuel consumption (mmbtu)	4 500 000	564 465[1]	1 012.3[3]		
Senior debt service coverage ratio base case	1.44 (1.31)	1.59 (1.23)	1.40–1.50	1.58 (1.37)	

Key:
1. lbs
2. Million lbs
3. Tons

Table 7.2   Comparison of five US power projects (continued)

Panel B: Principal project risks		
Carson Ice-Gen	*Colver*	*Indiantown*
1 Project-dependent revenue and cash flow stream: payments to creditors depend uniquely on project performance and output. 2 Uncertain impact of state regulatory liberalisation. 3 Declining costs for electrical capacity and energy, which may make the project burdensome to the power purchaser, as project power cost economies become less favourable. 4 Highly leveraged, resulting in low coverages. 5 Weak legal protection on excess cash flow distributions.	1 Unresolved post-construction problems with the EPC contractor over warranty issues, including structural and corrosion problems with the stack and high angle coal conveyor. 2 Operating risks of circulating fluidised bed boilers burning waste coal. 3 Above-market energy tariffs that could face future competitive pressures.	1 Low actual debt service coverage ratios.

Panel C: Technology		
Carson Ice-Gen	*Colver*	*Indiantown*
The LM6000 GE turbine does not have a long operating history, but even so appears to pose only moderate technical risk. While the 95% availability factor predicted by the independent engineer is not conservative, capacity payments are not reduced until the availability factor falls below 90%.	Operating performance indicates that the project's circulating fluidised bed boiler technology and its fuel handling and management procedures are proving to be a reliable power generation technology over the longer term. The unit is capable of handling fuels with heating values ranging from 6800–9500 Btu/lb and moisture contents of 5–15% with no problems.	Technology risk is deemed to be low, as the project depends upon known and proven technology. The O&M agreement contains incentives and liquidated damage penalties to ensure cost-efficient operation.

Panel B: Principal project risks

Rosemary	Brandywine
1 Dispatch risk as the plant is fully dispatchable.	1 Dispatch risk: the plant is subject to minimum dispatch requirements of 99 MW, but the remainder of capacity is fully dispatchable.
2 Potential loss of qualifying facility status, while the thermal offtake contract has no minimum take provision.	2 High cost of power to purchaser. The average all-in cost of power to purchaser over the life of the PPA is 10.5 cents/kWh, based upon 46% average dispatch.
3 Regulatory disallowance provisions in the power purchase agreement (PPA) pose risks to creditors.	3 The fuel contract expires prior to the end of the term of the PPA.
4 Mismatched energy payments, with the project's heat rate not being as low as the heat rate targets in the PPA.	
5 The fuel contract expires prior to expiration of the PPA.	

Panel C: Technology

Rosemary	Brandywine
Technology risk is low since the plant uses the commercially proven GE Frame 7EA turbine and Frame 6 turbine. Both can run on natural gas and fuel oil and both are capable of online fuel changes. A potential fuel switch accordingly would not cause an outage.	Technology risk is low owing to the fact the plant relies upon proven technology and was constructed more or less as designed. The plant consists of two GE Frame 7EA combustion turbines, with a guaranteed net dependable capacity of 230 MW and a heat rate of 7124 Btu/kWh.
The plant's heat rate compares unfavourably the PPA's contractual heat rate of 8900 Btu/kWh. Given the plant's variable dispatch pattern, fuel cost estimates based upon the contractual heat rate would be aggressive.	
The plant began commercial operations in 1990 and since then has had a history of forced outages. At no point, however, has the project exceeded the 25 day forced outage allowance included in the PPA.	

Table 7.2   Comparison of five US power projects (continued)

Panel D: Power purchase agreement		
Carson Ice-Gen	*Colver*	*Indiantown*
See above.	25 year power and capacity sales agreement to Penelec, under a non-dispatchable, power sales contract that is subject to curtailment for only 400 hours per year in the event of emergency loading. Performance standards require project annually to deliver at least 85% of its previous 3 year average on-peak capacity to avoid penalty payments to Penelec. Colver's capacity factor for 1986 was 85% with an availability of 97% for the year.	Project revenues under the PPA consist of monthly capacity and energy payments. The fixed capacity payment is based upon the 300 MW capacity multiplied by the sum of the capacity rate and the fixed O&M amount. Capacity rates start at $23 000/MW/ month declining to $10 000/MW/month after 20 years.
	Penelec can assign Colver's power sales contracts to any of its affiliates, subject to certain consents and conditions.	The fixed O&M payment of $5170/ month escalates quarterly at the rate of inflation. The capacity payment is subject to performance and based on availability. Availability is determined in keeping with a capacity billing factor (CBF) equal to the annual capacity factor plus or minus half of the difference between the on-peak capacity factor and 93%. Projected capacity factors for the CBF will be slightly higher than the annual capacity factor.
	Colver's sales tariffs have consistently exceeded average market prices, but even so are not out of line with other must-run QF projects. Continued high power costs invite the risk of greater of competition with an attendant negative impact on Colver's performance. Penelec is among the State's lowest cost producers, averaging about 3.5 cents/kWh. This compares with purchase costs from Colver of c. 5.95 cents/kWh for both on- and off-peak sales.	If the CBF is in the range of 87–92%, the project will receive full payment. For every 1% increase above 92%, a 2% bonus is conceded, up to a maximum of 10%. As the CBF decreases, a penalty in the same proportion is levied down to a floor of 55%. At 55% CBF, the purchaser can withhold capacity payments. A 55% CBF for 27 months could trigger termination under the power contract.
		There is little risk in the energy rate margin as the energy pricing in the power contract is virtually matched by the fuel supply pricing. The energy payment is based upon energy produced, priced at a base of $21.53/ MWh in 1992, adjusted by an applicable hourly efficiency factor, escalated quarterly in keeping with domestic coal price changes.

Panel D: Power purchase agreement	
Rosemary	*Brandywine*

Power costs to VEPCO are deemed economically marginal cost basis. The capacity payment structure is, however, front loaded and declines continuously over the life of the PPA. The capacity payment schedule is as follows:
1997–98: $11.65/kWh/mo;
1999–2005: $10.82/kWh/mo;
2006–2015: $8.32/kWh/mo.
Through 1998 PPA capacity revenues are high and dispatch low, while just the reverse applies over the balance of the contract. In effect, average power costs decline from 13.9 cents/kWh through 1998 to 7.4 cents/kWh, relatively economical for a peaking facility, through 2015, with a concomitant negative impact on project revenues.

Project sell power to PEPCO pursuant to a PPA that expires in 2021. PEPCO has the right to dispatch the plant based upon economic criteria, subject to a minimum dispatch of 99 MW on weekdays from 0800 to 2000 hours. Minimum dispatch requirements provide some measure of cash flow protection.

The PPA provides for capacity and energy payments, the latter based upon a combination of fixed and spot prices for natural gas that are well matched with the fuel supply arrangement and the plant's thermal efficiency. Fixed capacity payments are subject to maintenance of an 88% equivalent availability and range from $13.74/kW/month to $23.63/kW/month.

Capacity payments are subject to a series of adjustments, two of which are a matter of dispute between the plant and PEPCO. The first of these is based on the financing cost of the facility, which both sides claim is unspecified in the PPA. PEPCO wants to base the adjustment on the lease closing date of 12/96 and using a 12 year treasury bond rate of 6.36%; the project sponsors want to base the adjustment on the lease commitment date , that is, roughly two years earlier, and apply a T-bond rate of 7.94%.

The second involves the terms of a previously agreed reduction to capacity payments. Capacity payments can be reduced depending upon when PEPCO's peak load reaches 5697 MW. If the peak load occurs prior to 1999, the NPV of capacity revenue reductions is $14.8 million; if the threshold peak load occurs post-1999, the NPV of capacity revenue reduction rises to $24.8 million.

QF status is maintained through steam sales to a wholly owned subsidiary of the project and is used to generate distilled water in amounts sufficient to maintain QF status.

Power costs are high, averaging 10.5 cents/kWh. Capacity payments increase over time with the result that dispatch declines over the life of the PPA. For the first decade, the all-in cost of power to the purchaser (7.7 cents/kWh) is reasonably economical for a dispatchable plant. This is based on average dispatch of 46%. Thereafter, project becomes increasingly expensive to PEPCO with the result average dispatch declines 8 percentage points while the all-in power cost rises 12.3 cents/kWh.

Table 7.2 Comparison of five US power projects (continued)

Panel E: Fuel supply and cost		
Carson Ice-Gen	Colver	Indiantown
Capacity costs under the PPA approximate to $150/kWh and escalate at around the inflation rate. SMUD is also obligated to pay 1.15 × the actual energy costs of the project. The cost of energy under the PPA is c. 31 mills/kWh under probable dispatch assumptions. While these costs are lower than expiring power purchase contracts with SCE and PGE, they are still higher than the current regional cost of power.	Fuel and price risk is minimal. The project actively purchases its fuel requirements on the spot market and from mine owners at below pro forma costs. Colver has also been exploring its own fuel pile and claims that more fuel exists and at higher qualities than was originally believed. The project accordingly should have adequate fuel supplies for the duration of its PPA.	Current as available projections are lower than the price specified in the PPA for energy delivered. The all-in cost of power in 1996 amounted to 8.5 cents/kWh. availability risks are low. The project has a long-term coal supply contract that supplies up to 1.2 million tons per year, the maximum requirement, for 30 years. There is no minimum take under the contract. Existing supply contracts provide for 100% of the project's coal requirement and are backed by reserves of more than 30 million tons. The 1992 contract price is $26/ton, with an escalator that reflects changes in domestic coal prices to the purchaser of project output.

appears reliable enough although it is less clear whether such success reflects plant technology in isolation or in combination with the fuel handling and management procedures adopted by the facility's operators. However that may be, the plant appears to have fully vindicated sponsor expectations in respect of fuel choice. The facility is meant to burn coal of a low-power ranking ('waste coal') and appears able to handle coals with variable heat and moisture contents without any difficulty.

Technical risk in the Carson Ice-Gen project derives from the use of a relatively new generating unit, in this case the LM 6000 turbine produced by General Electric. The plant is estimated to have a 95 per cent availability factor, perhaps too aggressive in view of the uncertain performance of the generating unit. The issue is, however, of more than academic interest since the capacity payment to be received by the project, which effectively covers the plant's total fixed costs, depends upon the facility being able to realize a minimum 90 per cent availability factor. Whether there is enough cushion in the margin of difference between estimated and minimum acceptable availability is an open question.

Panel E: Fuel supply and cost	
Rosemary	Brandywine
The project has a long-term fuel management services contract with the Natural Gas Clearinghouse (NGH). The contract has a minimum take sufficient to satisfy the plant's fuel requirements to meet steam and chilled water sales. Each month the plant nominates the minimum daily quantity up to a maximum of 3075 mmBtu/day. NGC has a firm obligation to deliver the minimum requirement, with the price paid linked to the monthly Gulf Coast spot index. Additionally, NGC and the plant have an agreement to supply spot gas or fuel oil to satisfy the facility's dispatch obligations. The pricing is based on actual fuel acquisition costs.	Fuel supply arrangements are matched to energy payments. Gas purchases are made under three contracts all of which have 25 year terms. The supply contracts are not coterminous with the PPA. The plant has the capability to burn both natural gas and fuel oil and to change sources without disrupting generation. On-site storage of fuel oil amounts to 6 days of plant operation; an additional 1 million gallons is stored off-site. Any other fuel need could be purchased spot.
The fuel management contract expires in 2005, ten years before the PPA. There is thus some risk that fuel supply and transportation arrangements will not be in place when the NGC contract expires or that the new contract will not match the terms of the PPA.	

Turning now to the power purchase agreements, the first point worth noting is the considerable variability that applies across different projects (Panel D, Table 7.2). This in turn is largely a function of the pattern of *state* regulation. Being sovereign entities, state authorities in the United States have wide latitude to set applicable guidelines in keeping with local requirements. We leave aside for the present the Carson Ice-Gen project since the development of independent power production represents an attempt by the sponsoring utility, the Sacramento Municipal Utility District, to reduce historic purchased power costs. From the standpoint of project revenues the critical issues are the project's ability to dispatch power at acceptable cost – something that may cause the purchasing utility to meet its requirements relying on other sources – the terms governing the tariff structure, and whether the purchaser has an obligation to purchase a minimum quantity of power if the facility is able to supply it.

For the four remaining projects shown in Table 7.2, project tariffs contain a capacity payment, set either as a fixed or variable charge. The fixed capacity payment for the Indiantown project, for example, is based upon the plant's 300 MW capacity multiplied by the sum of the capacity rate and the fixed operating and maintenance amount. The fixed capacity payment, initially set at \$23 000/MW/month, declines over project life, in keeping with the repayment of the project's outstanding debt, to around \$10 000/ MW/month after two decades. Note, finally, that the capacity payment is subject to both performance and availability, with the latter determined according to a formula that sums the annual capacity factor plus (or minus) one half of the difference between the on-peak capacity factor and 93 per cent. The project receives full payment if availability is in the 87–92 per cent range; the plant earns a 2 per cent bonus for each 1 per cent above base availability (93 per cent) but capped at 10 per cent. On the other hand, declining availability can result in the imposition of financial penalty. In the event that plant availability declines to 55 per cent or lower, the purchaser has a contractual right to withhold capacity payments. If availability remains at or below the 55 per cent level for more than two years, the purchase agreement could be terminated.

Similar features are contained in the power purchase agreements applying to the other projects surveyed. Capacity and purchase agreements are typically of predetermined duration – generally around 20 years – while capacity payments tend to decline over time. For the Rosemary project, for example, capacity payments decline by about 30 per cent over contract life, from \$11.65/kWh/month in 1997–98 to \$8.3/kWh/month for the period 2006–15. The Brandywine project illustrates some of the difficulties that may arise over project life in connection with determination of capacity payments and the decline curve applying to such payments. It also demonstrates the risks creditors face when the terms and conditions governing key contract provisions are left unspecified.

In this case, capacity adjustment payments include the financing cost of the facility. Project sponsors contend that the capacity payment depends upon the date the lease commitment was entered into, and not the closing date (some two years later) as claimed by the purchaser. Similarly, plant owners favour using a US Treasury bond rate of 7.94 per cent as against the 6.36 per cent rate favoured by the utility. The second disputed issue relates to the date at which at capacity payment reductions can be implemented. The utility company alleges that reductions are applicable if the peak load threshold is achieved prior to 1999; project sponsors, by contrast, argue the relevant date is after 1999. Nor are these conflicted interpretations inconsequential: depending upon which cut-off date is chosen, the difference in the present value of future payments to Brandywine project owners amounts to \$10 million.

We might note, finally, that in the case of the Colver, Rosemary and Brandywine projects, power costs tend to be high compared with either total purchase costs or own generating costs. In the case of Colver, for example, plant generating costs exceed Penelec's (the utility committed to purchasing the project's power output) average purchase price by almost 2.5 cents/kWh, or a premium amounting to around 70 per cent These considerations could result in both low dispatch and capacity factors, with an attendant negative impact on project cash flow. Of greater importance is the relationship between capacity payments and capacity utilisation. For the Rosemary project, capacity payments decline to nearly one half of their original level by 2015. True, dispatch increases, but it is less clear whether the offsetting impacts result in higher project revenues than would otherwise have occurred. For the Brandywine project, just the opposite applies: capacity payments increase more or less steadily but at the cost of progressively lower dispatch, in fact by some 8 percentage points over the second half of the term of the power purchase agreement.

Fuel supply contracts and accordingly energy payments are as variable as are power purchase agreements (Panel E, Table 7.2). For the five projects under review two burn coal as the primary fuel input, while the others use natural gas; the Brandywine facility, as noted, is dual fuelled, meaning that it can source, depending upon prevailing market prices, either natural gas or fuel oil to generate electricity. Each project also covers its fuel requirements under different contract structures. Colver, for example, relies primarily upon spot purchases but claims to have substantial coal reserves of generally high quality in its own fuel pile. The Indiantown project, by contrast, has a long-term coal supply agreement that provides for up to 1.2 million tons per year, equal to the plant's maximum fuel requirement. The contract is backed up by reserves that equate to 25 years of plant fuel consumption and was set initially at $26/ton. Prices escalate in keeping with the cost of coal to the purchaser of plant output.

Natural gas supplies for the Rosemary and Brandywine projects are also covered under long-term supply agreements. In the case of the former facility, the supply contract expires 10 years before the power purchase agreement, exposing creditors to both availability and transportation risks. The agreement under which Rosemary sources its natural gas provides for a minimum take sufficient to satisfy the facility's requirements to meet power and chilled water sales. Volumes up to a pre-set ceiling are nominated monthly by the plant, with the supplier legally obligated to deliver such quantities. Prices are linked to a monthly US Gulf Coast spot gas index. Fuel supply requirements for the Brandywine plant are covered under three supply agreements, all of which have terms of 25 years. The facility also has on-site fuel oil storage equal to 6 days plant operation; an additional 1 million gallons is stored off-site. Nor is the plant likely to face any difficulties obtaining fuel oil supplies on the spot market.

The transition from independent to merchant power production increases the uncertainty of project cash flow and commensurately the willingness of lenders to provide funds on terms similar to those that applied historically. The most obvious change is that producers no longer enjoy tariff structures that effectively guarantee that the project will be able to cover its fixed and variable costs, including an appropriate rate of return to the investment's sponsors. The financial interests of creditors were similarly safeguarded, since the authorised price structure generally included debt service in the definition of fixed charges. Second, projects no longer have an automatic outlet for surplus power production. In merchant projects, electricity output is sold into a national or transnational power pool where prices can fluctuate widely in keeping with time of day or seasonal pressures.[9]

To many commentators, production costs are the *sine qua non* for successful merchant projects. Low-cost producers will be able to dispatch more electricity with an attendant favourable impact on plant capacity factors. Project cash flow and plant capacity factors are, quite obviously, positively correlated as is the profitability of the facility. Indeed, numerous commentators stress the importance of the merchant plant's position on the industry cost curve as a lead indicator of potential financial performance. The reality, however, is that most new plants will be more or less similarly positioned. Unless small differences in cost structures or management ability translate to disproportionately large financial benefits, a not entirely unrealistic assumption as we know from other industrial sectors, then the prospective market balance is bound to become the more critical determinant of project loan values.

Competition, however, is not likely to be confined uniquely to plants constructed more or less at the same time, on the same scale and employing basically the same technology. Indeed, one of the consequences of greater competition within the electricity supply industry is that existing utilities are now in the process of divesting their generating assets, concentrating instead on marketing and transmission. Generating companies whose costs have been fully amortized are clearly at a competitive advantage *vis-à-vis* greenfield projects. One way merchant plants could correct the imbalance, all things given, would be to replicate the capitalisation of established plants. This cost may be avoided if, as could easily be the case, the merchant plant is technically more advanced than existing generating units. In effect, the financial advantages enjoyed by existing generating plants could be offset by greater cost and operating efficiencies associated with state of the art technology.[10]

Otherwise, project sponsors must be prepared to inject more capital into greenfield merchant projects than they would prefer and, as a corollary, accept that lenders are unlikely to provide debt in the same amounts or on the same terms as previously. In this latter connection, we might note that lenders will require higher coverage ratios; they will also require merchant plants to maintain larger cash reserves, of the order of 1 year or longer, to cover both debt service risks and operating and maintenance charges.

Moreover, these accounts will have to be fully funded prior to the commencement of plant operation. Collectively, these considerations would appear to seriously undermine the rationale for choosing limited recourse debt over alternative financial options.

One way in which regulatory authorities might level the playing field would be to allocate to divested generating companies the environmental liabilities of their former parent companies. Whether this neutralises the financial benefits of fully amortised plants will obviously vary from generator to generator; at the extremes it could enhance the competitive position of new entrants, which might theoretically enjoy lower overall costs. On the other hand, regulatory authorities could just as easily, and with perhaps greater justification, require the utility company's shareholders to absorb such liabilities, resulting in the loss of a potentially sizable financial offset to greenfield projects. However the balance is ultimately struck, the current pattern, and the responsiveness, of regulation to changing market conditions is one of the principal ongoing risks faced by project developers and, by extension, lenders.

Nor can either take much comfort from the historical pattern of regulation. However much regulators may want to develop policy independent of immediate political pressures, the simple fact is that as agents of the legislature that created them they actually have little room for manoeuvre. Even if they did, sooner or later political realities would catch them up. In the United Kingdom, for example, the period within which tariffs are now reviewed has been shortened. Moreover, the government recently enacted a 'one time' windfall profits tax, while the so-called 'dash for gas', instrumental in supporting a high level of independent power development, has been abandoned in favour of a new policy aimed at enhancing the position of the domestic coal industry in the electricity generating market.

At a stroke, the British government simultaneously altered national environmental and fuel priorities. In more normal circumstances one might have expected the opposite political reaction – a tightening of existing environmental rules by, for example, the adoption of policies favouring natural gas over coal or fuel oil. Either way, recent British experience underscores the critical importance of the vagaries of regulation (more properly, political decision making) as a key project risk in either independent or merchant investments.

The range of risks to be faced by lenders on the demand side of the market is equally formidable. The ideal would be for the market in which the project is being developed to be growing rapidly, faster in fact then potential generating capacity. Lenders are bound to view projects where buyers have considerable control over prices, either directly (through the scale of purchase commitments) or indirectly (via a regulatory regime that favours consumers over generators), less positively than those where the opposite applies. In the same vein, lender interest in a given project might be expected to vary inversely with the degree to which prospective purchasers are able to switch suppliers. As the degree of fuel switching is a function of the underlying market balance, this reinforces the point made above concerning the importance that financial markets attach to near- and medium-term market prospects.

More positively, the transition to more volatile trading conditions could be a source of strength for those sponsors best able to manage to price risk. Lenders are familiar with the ways hedging strategies can be integrated into project financings, so that structures similar to those used in traditional commodity financings can be implemented and to the same effect. In other words, lenders should explore the ways commodity risks, now firmly established in the power industry, can be mitigated or eliminated rather than to try to identify projects that fill unique market niches. Again the same concerns heard following the 1986 oil price collapse, that volatility is detrimental to future industry growth, are being repeated in the power market. Increased price volatility has had no lasting negative impact on petroleum industry investment. The same is bound to be true of the power market. Instead of lenders and utility companies bemoaning the loss of 'stability' or searching for ways it could be recreated, both would be better advised to concentrate on the opportunities that follow from market liberalisation.

TRANSPORTATION PROJECTS

We previously alluded to the fact that transportation or transport-related projects are more likely to be cross-border than are most other infrastructure investments. In this narrow sense such investments closely resemble more traditional project financings. Still, we should not lose sight of the fact transport projects can and do present risks that are fundamentally different, both within a given market as well as across different transport sectors, from those applying to, say, natural resource investments. Airport projects, ostensibly the most attractive of all transport sector investments, well illustrate the significance of these differences.

The broad receptivity of financial markets to airport investments stems mainly from two unconnected considerations. First of all, user charges for international traffic are denominated in hard currency, typically US dollars. Lenders and sponsors, accordingly, have immediate access to foreign exchange that can be used for debt service and dividend distribution, respectively. And second, traffic – and by extension revenues – tend to show more or less steady and predictable growth. The implied positive impact on both the level and stability of potential cash flow means that lenders are normally prepared to concede more favourable coverage ratios than apply to projects in more traditional sectors. On the other hand, it is worth emphasising that airport assets are not homogenous, especially in respect of economic life. Runways, taxiways and aprons, which if properly maintained, can last as long as 20–30 years; passenger and freight terminals, by contrast, have shorter economic lives, of the order of 10–20 years. Owing to these longevity differentials, lenders will normally evaluate each asset class separately and apply funding structures appropriate to the risk characteristics of the different segments.

Private funding of airport investments is less common than might have been expected. As a general matter this has less to do with a lack of sponsor or lender appetite for such investments, than with circumstances more or less unique to the sector, which unnecessarily complicate the application of limited recourse debt structures. In many countries, above all, in emerging market economies, no distinction exists between civil and military aviation; the two are often one and the same. Given the lack of such distinctions, airports and related infrastructure tend to be considered strategic assets, making it difficult, if not impossible, for private foreigners, alone or in combination with local investors, to manage and control local air traffic facilities.[11] The same arguments, of course, do not apply to the management of other airport facilities. Investors looking for airport investments would, accordingly, be better advised to concentrate on these latter activities.

Compania De Dessarollo Aeropuerto Eldorado S.A. (CODAD) clearly illustrates the principles applying to private sector management of airport facilities and the requirements that must be satisfied if international project finance markets are to provide funds for project development (Simonson and Haddad, 1997). The project involves construction of a second runway at Colombia's Eldorado Airport, located just outside of Bogota, the country's capital, as well as maintenance of both runways over the life of the concession, currently scheduled to run until 30 September 2015. In return for such services, from the time the second runway is operational until the concession expires CODAD will have the right to collect landing fees generated by commercial carriers.[12] The economic justification for the project derives from the fact that the existing runway is operating at or near full capacity. Total enplanements have increased by 6 per cent annually since the mid-1980s, with most of this growth having occurred in 1991–94. Authorities have dealt with the resulting increased congestion by promoting off-peak schedules and tolerating operating rates that exceed traditional safety norms. Given that the two runways will lie parallel to each other, both are expected to operate concurrently.

From a lender's standpoint the critical project issues are: (1) what mechanisms exist to protect creditors in the event construction is delayed or the contract is cancelled, (2) how robust are prospective air traffic growth forecasts, (3) what are the tariff and payment mechanisms applicable to the project and (4) are there any safeguards built into the concession that protect creditors in the event revenues do not correspond to forecast levels?

One of the issues central to lender interest in the project is the fact that it is of critical importance to the Colombian economy, intended as it is to create the additional capacity needed to accommodate projected rapid air transport growth. More encouraging still is the expectation that the growth of international emplanements will continue to exceed domestic transport market growth, although over the concession period the margin of difference is expected to compress significantly. Between 1986 and 1994, international emplanements increased by 8.4 per cent per annum, 3 percentage points per year faster than domestic growth. According to recent projections, international emplanements are forecast to increase by 4.1 per cent per annum versus 3.9 per cent for domestic growth, slashing the growth differential to a mere 25 basis points per year. Even so, by the end of the century, international operations will be 1.5 times higher than a decade earlier.

The effects of technical or financial difficulties arising during the construction phase of the project are handled in several ways. If such problems originate with the contractor, the resulting losses may be covered by penalty payments made to CODAD by Aerocivil, the national aviation authority, equivalent to America's Federal Aviation Agency. The construction contract also contains liquidated damage provisions, meaning that Dragados and Conconcreto, jointly and severally, will have to pay CODAD up to $42 million, an amount equal to two-fifths of the engineering placement contract (EPC). In such circumstances, Dragados is also obligated to purchase the outstanding debt. If, by contrast, construction is halted owing to actions taken by national authorities, then either the construction period can be extended or, if the problem persists, CODAD has a right of termination. If the latter option is exercised, Aerocivil is legally bound to pay CODAD an amount sufficient to redeem the debt.

Once construction is completed, CODAD has the right to collect landing fees generated by both domestic and international carriers. Roughly 70 per cent of total revenues are expected to derive from international flights, providing creditors with a significant hedge against currency risk. Rather surprising, all revenues are payable in local currency. Fees are fully indexed, to the US Consumer Price Index (CPI) and the local retail price index for international and domestic flights, respectively, and adjusted semi-annually. Any increase exceeding 5 per cent in the case of the US CPI or 10 per cent in the local index automatically triggers immediate upward adjustment. Tariffs can, however, be adjusted at any time by Aerocivil.

Landing fees are paid into an *onshore* account, with disbursement priorities dictated by the structure of the financing agreement.[13] An additional safety feature is a minimum revenue guarantee, an unconditional commitment on the part of Aerocivil to cover any difference between actual and mutually agreed minimum project revenues. In the event of a shortfall, payments are to be made semi-annually from a trust account established by

Aerocivil. The guarantee is denominated in US dollars, with the account funded in an amount equal to 30 per cent of the minimum revenue guarantee. These funds can also be used to cover tariff modifications as may be required by law, unforeseen environmental costs and regulatory or income tax law changes.

Over the life of the original concession, debt service coverage ratios are favourable and increase more or less steadily, from around 1.6× in 1998 to 2.8× in 2011. The steady improvement in coverage is a function of three variables: (1) forecast higher air traffic, (2) increases in landing fees, and (3) unchanged real (that is, inflation-adjusted) operating costs. The principal economic assumptions underpinning the cash flow projections are that the dollar/peso rate will follow purchasing power parity, that is, the Colombian currency will depreciate by no more than the difference in the local rate of inflation *vis-à-vis* the United States. Colombian inflation is expected to recede from the current 18 per cent annual rate to 10 per cent by 2004; the US rate of inflation, by contrast, is assumed to be a constant 3.8 per cent per annum. Table 7.3 summarises the key financial projections.

Maritime ports, too, comprise different asset classes each having their own unique characteristics. Jetties and wharves, for example, again even without regular or careful maintenance, are likely to last for up to fifty years or longer. Handling and storage equipment and warehouses, by contrast, have shorter economic lives, generally amounting to no more than 10 years. Again this means that lenders will require separate funding sources to be applied to each market segment. Generally speaking, it is easier for financial markets to match tenors to storage and handling equipment financings than

Table 7.3 Eldorado International Airport base case financial results and debt service coverage (thousands of US dollars)

Year	Operating revenue	Operating expenses, taxes and fees	Net income	Debt service	Coverage ratio
1998	10 858	1 554	9 304	5 828	1.60
1999	27 263	5 386	21 877	14 687	1.49
2000	29 089	5 877	23 932	15 748	1.52
2001	31 597	6 376	25 221	16 648	1.51
2002	33.593	7 064	26 529	17 418	1.52
2003	35 679	7 644	28 035	18 027	1.56
2004	37 734	8 304	29 430	17 673	1.67
2005	40 280	9 365	30 915	18 360	1.68
2006	42 785	10 308	32 477	17 491	1.86

Source: Simonson and Haddad (1997).

to wharves or jetties, since the economic life of the latter exceed by a wide margin the longest tenors applicable to either commercial bank financings or project bonds. Moreover, lenders are likely to be more interested in certain port-related activities than others. Ports by definition are extremely diverse, with numerous activities being carried out simultaneously some of which, freight forwarding and loading, for example, being more suitable for private sector management and control.

In the same way we might draw a clear distinction between greenfield port developments and privatisations, the former confronting investors with a wide range of difficult, even intractable, issues. Nor does privatisation of existing port operations necessarily obviate these sorts of problems. The most attractive privatisation opportunities appear to be those where the host government is willing to treat previous investments in jetties and wharves as a sunk cost. Such investments are now properly seen as involving the modernisation of port operations and administration rather than an attempt at recovering past development costs.

There are several other features of port developments that are a source of ongoing concern to financial markets, hence explaining the dearth of such issues. First of all, ports are labour intensive and more often than not are under the direct control of trade unions. This combination of circumstances creates the objective conditions upon which host government and investor interests coincide, namely, the desire to increase productivity by eliminating wasteful rules that by design are intended to maximise employment. The successful elimination of such practices creates significant profit potential; it is also typically associated with redundancies that could, depending upon their magnitude, have significant financial and political ramifications.[14]

Unless labour relations can be sorted out satisfactorily in advance, the prospects for successful private sector management are minute. Indeed, it is arguable whether any private investors would have the least interest in port operations unless these difficulties were sorted out prior to the award of the concession This issue takes on even greater significance once we recall that port revenues are notoriously difficult to predict, owing mainly to the short-term orientation of shipping companies and the risk that international trade patterns can change abruptly as they have done. Still, it is possible to cite examples that prove ports can be developed successfully as private concessions. Blaicklock (1998), for example, argues that the most successful port developments will usually involve the combination of dedicated terminals, relatively low labour costs and operations that embrace state of the art loading and storage technologies; iron ore and coal terminals, for example, fall into this category.

Private road projects are another area of infrastructure development fraught with problems. It is, for example, extremely difficult to quantify potential usage and by extension project revenues unless the investment has monopoly or quasi-monopoly characteristics, tunnels or bridges, for example. Failing that, investors would need to have reasonably precise estimates of the elasticity of traffic demand both to prospective user charges and income growth. Income elasticities of demand are bound be quite high, since the growth of GDP affects both discretionary road usage, via its impact on the demand for leisure travel, and on freight volumes, in keeping with the demand for intermediate production inputs and final outputs. Price elasticities might be expected to be higher the greater the number of alternative routes available to potential users. Unless these considerations are properly factored in the analysis, the result invariably will be wildly optimistic traffic projections that invite project failure.

Nor, as might be expected, is this problem unique to road investments in emerging market economies. The classic examples here are the M1–M15 toll road linking Budapest and Vienna where low usage, well below forecast levels, resulted in swift financial default. Other examples include the Lyon Ringroad, where similar difficulties resulted in the French government having to buy back the project from the original sponsors. The Hardy Toll Road in Houston, Texas, linking the city centre and the airport provides yet a third instance of the same phenomenon. Project sponsors appear to have grossly overestimated user perceptions of the opportunity cost of time, theoretically one of the main inputs into the derivation of the 'correct' toll. In consequence road usage is well below initial expectations, with business travellers by and large the principal users.[15] Lastly, road projects present more significant environmental risks than do most other infrastructure investments. Unless properly addressed, such risks could easily frustrate smooth project realisation, imposing substantial additional costs on the investment to the detriment of prospective returns.

Against this backdrop it seems odd that traffic forecasts still show fairly narrow confidence bands. Consider, for example, traffic projections for the five Chinese expressway concessions developed or under construction by Greater Beijing First Expressway, Ltd (GBFE) (Doud and Forsgren, 1997). The first panel of Table 7.4 presents summary characteristics of the concessions held by GBFE, while the lower panel shows annual traffic projections compiled in March 1997 by a Hong Kong-based consultancy with direct knowledge of China's transport sector. To place these projections in perspective, passenger volumes on all Chinese roads rose by 11.7 per cent per year between 1991 and 1995, while freight volumes were up by 6.5 per cent over the same period. Moreover, Chinese road densities, whether measured in relation to total population or land area, are among the lowest in the world. On either index, China falls well below other Asian economies being only 61 per cent and 17 per cent, respectively, of the comparable South Korean figures. Even in comparison with India, almost certainly at a lower level of development than China, the results are still unfavourable, with densities relative to population amounting to only 40 per cent of the Indian figure and 17 per cent in relation to area.

Table 7.4   Summary of key features of five GBFE Chinese expressways

Key project characteristics					
	Zhoubao	*Baoshi*	*Tianjin South*	*Hebei*	*Tianjin*
Length (km)	105.7	116.3	41.8	21.2	37.1
Lanes	4	4	4	4	4
JV start date	3/97	9/96	8/96	3/97	9/96
Total JV period	21	21	22.5	28	28
Year of operation	1994	1994	1995	2000	2000
Speed limit	120	120	100	120	120
Total cost[1]	$181	181	115	87	133
Cost/km	$1712	$1556.3	$2751.2	$4103.8	$3584.9
Toll rate[2]	26–86	26–86	33–78	40–150	40–150

Traffic projections: average annual growth rate, annual daily traffic forecasts, 1997–2020

Base case	7.6	7.3	5.7	6.1	6.7
Conservative case[3]	7.8	7.5	5.7	6.7	7.2

Source: Doud and Forsgren (1997).

Notes:
1 Millions of US dollars.
2 Fen per kilometre.
3 Even though initial and terminal values are lower in the conservative than in the
   alternative scenario, base level effects account for the indicated faster rate of increase.

These indices provide what appears to be a wide margin of comfort in respect of potential demand growth and by extension the project's ability to satisfy base case debt coverage ratios. These perceptions are strengthened once it is recognized that most secondary roads are in poor condition and heavily congested. And lastly, the new expressways can be used by heavy goods vehicles, coal trucks, for example, which previously would have been banned. On the other hand, the projections are based upon what in hindsight appear to be extremely optimistic assumptions regarding near- and longer-term economic and financial trends. Sensitivity analysis covering a wide range of risks, alone or in combination, indicates that project economics appear robust enough to 'withstand significant downside factors without jeopardizing both interest and sinking fund payments. For example, one sensitivity case included only 50% of the base case toll increases coupled with *renminbi* devaluation, completion problems and 5% traffic growth 2000 and 50% of the projected growth thereafter' (Doud and Forsgren, 1997).

It is arguable, for example, whether project economics were ever evaluated taking account of the risk of a regional financial crisis or the significant revaluation of the renminbi *vis-à-vis* other regional currencies, rather than devaluation against the US dollar. While China now favours maintaining the current exchange against the US dollar, a policy backed up by substantial foreign exchange reserves and the apparent willingness of the authorities to deploy such resources in defence of the currency, it is questionable whether devaluation can be avoided. The combined impact of recent developments is to threaten the country's immediate and medium-term financial prospects, owing in large measure to the significant depressing effect recent exchange rate changes have had on China's international competitiveness, one of the main drivers of past rapid rates of growth. Slower growth translates to reduced road usage and, as Chinese authorities are presumably unwilling to add to antecedent cost pressures, we might expect growing resistance to future toll rises.[16] It is worth noting in this connection that the various concession agreements provide no specific adjustment mechanisms in respect of either local inflation or devaluation. The irony is that what in early 1997 was regarded as 'worst' case could easily be considered today as base case.[17]

The last transport sector we shall consider is railroads. Rail networks, like air and maritime ports, incorporate both long and short-term components. Tracks and stations constitute the principal long-term assets while equipment and rolling stock represent short-term assets. These differences again mean that lenders are likely to view the two asset classes distinctly and in consequence will assign loan values in keeping with strength of the commitments underpinning each. Railroads differ from alternative transport modes, above all, buses and trucks, in one critical respect, namely, they are highly capital intensive. Huge up front costs add to the difficulty of quantifying the financial merits of rail projects, especially in comparison with competing transport sectors. On the other hand, it has been argued that once environmental costs are factored into the analysis, the economics of rail operations often emerge as superior to other forms of transportation (Blaiklock, 1998). Rail networks confront lenders with the same difficulties forecasting passenger and freight volumes and revenues we encountered in other forms of transportation, and for the same reasons lenders will tend to evaluate project economics using extremely conservative criteria.

Rail networks are geographically dispersed making it extremely difficult precisely to define and limit the concession area, an ideal prerequisite of concessionaires, which thus know the full extent of their commitment; lenders, too, for rather different reasons prefer the concession to be fully carved out. The most successful rail network developments, accordingly, involve a dedicated end-user, while the most successful rail privatisations, like the most profitable port developments, involve transfer of the associated infrastructure, rights of way, tracks and stations at below fair market values.[18] In addition, developers will often seek to obtain the right to develop adjacent properties, either in support of the concession or as an independent source of project income.

The Eurotunnel project illustrates all of the pitfalls connected with developing a greenfield rail network, with ongoing operating results continuing to vindicate the most pessimistic assessments of the project's economic and financial merits. Total project cost amounted to roughly twice original estimates; operating revenues, too, are running below projections, notwithstanding reasonably accurate market penetration forecasts. Tables 7.5–7.7 present selective data on different aspects of the Eurotunnel project. By way of background, we might note that Eurotunnel comprises four separate train services. Tourist and freight vehicles under the Le Shuttle name operate between the UK terminal located near Folkestone and the French terminal at Coquelles. Eurostar passenger services and through freight services are operated by concessionaires that pay Eurotunnel for the right to use the tunnel. Eurotunnel generates additional revenues from the provision of duty-free, tax-free and tax-paid shopping and catering facilities at the terminals as well as ancillary services, such as telecommunications and property development.

The Le Shuttle tourist service transports mainly cars, coaches, minibuses and motorcycles between the English and French terminals. Shuttles comprise two sections: a double decked section, used principally by autos and motorcycles and a single deck section for coaches, minibuses and cars with caravans. Customers drive to the terminals and may use either the facilities located within the terminals or alternatively head directly for the first shuttle departure. Tickets can be purchased in advance from travel agents or on arrival at the toll stations. Vehicles are accompanied by their owners throughout the journey. Le Shuttle freight services carry heavy goods vehicles between English and French terminals, but unlike tourists, truck drivers and their passengers travel separately from their vehicles in a Club Car which provides catering services. Tickets are purchased on arrival at the terminals or, alternatively, rates can be negotiated directly between users and the carrier. Amenities are available for truck drivers at or near the terminals, including duty free shopping, while tolls, customs and security clearance are provided through dedicated facilities.

Eurostar employs high speed trains that carry passengers between London, Paris and Brussels. Fourteen or more scheduled services operate between London and Paris in either direction and eight per day between London and Brussels. On certain services, passengers can board or alight at intermediate stops, Ashford in Kent, Calais-Frethun or Lille. With regard to freight, three different services operate through the tunnel: (i) freight containers and swap bodies, which account for two-thirds of through trains; (ii) conventional rail wagons, which account for just under a fifth of traffic; and (iii) automotive traffic, comprising the transport of new vehicles and components, and accounting for 14 per cent of through traffic. Through freight services are principally intermodal operating between the United Kingdom and Italy, France, Spain and Belgium. Automotive trains carry

Table 7.5   Forecasts of Eurotunnel system total revenues and Eurotunnel annual passenger and freight movements, 1996–99

	1996	1997	1998	1999
Forecast total revenues (millions of 1996 £)[1]				
Original projections				
Le Shuttle	451.0	492.7	478.7	493.3
Rail	359.3	365.6	372.0	378.6
Other	75.7	77.9	80.0	82.3
Total	*886.0*	*936.2*	*930.7*	*954.2*
Revised Projections				
Le Shuttle	145	128.9	235.5	274.3
Rail	198	200.2	205.1	222.4
Other	140	224.6	175.7	109.4
Total	*483*	*553.7*	*616.3*	*606.1*
Forecast cross-Channel passenger and freight traffic, 1996–99				
Original projections				
Passenger[2]	90.4	93.3	96.3	99.4
Freight[3]	94.4	98.0	101.7	105.6
Revised Projections: Le Shuttle				
Passengers	5.33	5.34	5.77	5.55
Freight	1.28	1.50	1.63	1.77
Revised Projections: Eurostar				
Passenger[2]	89	95	101	104
Freight[3]	55	59	64	68
Market share (per cent)				
Original projections				
Passengers[2]	32.0	31.6	31.4	31.1
Freight[3]	17.3	17.2	17.2	17.0
Revised projections				
A Le Shuttle				
Car[4]	41	49	59	62
Coach[4]	29	38	45	45
Accompanied HGV[4]	41	42	56	57
B Eurostar				
Passenger[1]	5.5	7.1	9.5	10.2
Freight[2]	4.4	4.7	5.0	5.1

Source: Finnerty (1996), Eurotunnel (1997).

Notes:
1 Original revenue projections were derived by applying indicated growth rates to base 1993 figures and converting from constant 1987 to constant 1996 prices.
2 Millions of passenger trips per annum.
3 Millions of tonnes per annum.
4 Million vehicles.

Table 7.6   Eurotunnel economic and financial forecasts

		A. Medium term	
		Revised	
Variable	Original	1996–99	2000-06
Economic growth			
United Kingdom	2.15% per annum between 1985 and 2003; 2.0% p.a. through 2013	2.3% (96), 3.2% (97), 2.8% (98), 2.5% (99)	2.5% (00), 2.4% (01–03), 2.0% (04–06)
France	NA	1.2% (96), 2.4% (97), 2.1% (98), 2.1% (99)	2.1% (00–03), 2.0% (04–06)
FF:£ exchange rate	10:1	8.5:1	8.5:1
Inflation			
United Kingdom	6% per annum from 1991	2.4% (96), 2.9% (97), 3.4% (98), 3.6% (99)	3.9%(00), 4.0% (01), 4.1%(02), 4.3% (03–06)
France	NA	2.0% (96), 1.5% (97), 1.6% (98), 2.1% (99)	2.5% (00), 2.7% (01), 2.9% (02), 3.3% (03–06)
Interest rates	On cash balances: 8.5% per annum through concession period		
United Kingdom (Inter-bank)		5.7% (96), 6.1% (97), 7.0% (98), 7.4% (99)	7.5% (00), 7.6% (01–02), 7.8% (03), 7.2% (04), 6.5% (05), 5.9% (06)
France (Inter-bank)		3.9% (96), 4.1% (97), 4.6% (98), 5.4% (99)	5.7% (00) 6.0% (01), 6.2% (02), 6.3% (03), 6.0% (04), 5.8% (05), 5.5% (06)

Table 7.6 (continued)

Variable	Original	B. Extended projections Revised Upper case	Lower case
Total traffic (annual average growth rate)[1]			
Passengers	2.81%	5.7%	5.7%
Freight	3.56%	8.9%	8.1%
Eurotunnel (market share)			
Passengers	39% (03), 36% (13)		
Freight	17% (03), 16% (13)		
Le Shuttle			
Passengers		67% (03), 70% (06)	67% (03), 70% (06)
Freight		42% (03), 60% (06)	44% (03), 61% (06)
Eurostar			
Passengers		10.2% (03), 11.9% (06)	8.8% (03), 9.2% (06)
Freight		5.1% (03), 4.7% (06)	3.3% (03), 3.4% (06)
Eurotunnel revenue (average annual real growth rate)[1]			
Shuttle	2.3% p.a	13.2% p.a	11.5% p.a
Rail	1.3%	1.6%	−0.2%
Other	2.1%	−13.5%	−13.5%
Total	1.9%	4.7%	3.2%

Source: Finnerty (1996), Eurotunnel (1997).

Notes: Growth rates calculated from 1993–2013 in the original forecast and between 1996 and 2006 in the revised projection.

Table 7.7   Risk factors affecting Eurotunnel project

Factor	Risk
	Operating risks
HGV Fire	Caused significant damage to both a section of Tunnel and the Le Shuttle freight train concerned in the fire. All services through the tunnel were temporarily interrupted. The Directors are of the opinion that Group insurance policies will cover (net of deductibles) the cost of repairing damage to the tunnel concerned, replacing fixed equipment and rolling stock, and the loss of revenue for a period not exceeding one year. On the other hand, there is no assurance that such payments will fully cover such costs.
Capacity	Future revenue growth is dependent upon having sufficient capacity at the time to cope with the associated traffic flows. Any increases in the utilisation of the tunnel may be subject to Inter-Governmental Commission (IGC) approval, which may not be immediately forthcoming. A key factor determining system capacity is the signalling system that regulates traffic through the tunnel. The Group anticipates that enhancement of the signalling system will be necessary using technology which is not presently available commercially.
	Eurotunnel has agreed with the safety authority that Le Shuttle freight services should resume with three temporary amendments to the intended long-term operating arrangements. Although it is believed that these will have a significant impact on the capacity of the service, there can be no assurance that these temporary arrangements will be lifted.
Rolling stock reliability	Reliability of rolling stock, especially in relation to Le Shuttle freight, has been a source of continuing difficulty, owing partly to their design and manufacture. There can be no assurance that work to address these problems will be successful. Continuation of reliability problems will limit the capacity of Le Shuttle operation.
Inter-Governmental Commission	The IGC authorizes the services which operate through the tunnel; as such it effectively supervises the activities of Eurotunnel. Eurotunnel is free to operate the tunnel so long as it complies with all IGC requirements and the safety authority. The IGC can review and amend its requirements in respect of Eurotunnel's operations. The main circumstances under which the IGC might exercise such authority is: significant incidents which occur in the tunnel, such as the HGV fire; failure by Eurotunnel to comply with existing requirements of the IGC and safety authority; and general changes relating to the regulation of railway operations.
Extension of the concession	Extension would be beneficial as it would reduce the Group's annual depreciation charge, increase reported annual after-tax profit or annual after-tax loss, and bring forward the date when the first dividend payment may be made.

Table 7.7 (continued)

Factor	Risk
	Market risks
Competing services	Significant rationalisation of ferry services that could have significant competitive implications. So too could new modes of transport or infrastructure in the markets in which Eurotunnel operates. For example, high speed ferries on cross-channel routes and the establishment of a second fixed link by another operator if Eurotunnel does not build one.
Duty-free	The duty-free regime within the EU is to be abolished on 30 June 1999. Cross channel operators derive considerable revenues from such sales. On the other hand, extension of the regime would obviously extend the period over which such revenues could be earned, with the result of delaying the anticipated recovery in yields for cross channel crossings.
EU railway infrastructure directives	EC rulings provide, *inter alia*, rights of access to and transit through national railway networks. An open question is how these rules apply to Eurotunnel.
	In 1994, the EC decided to grant the Railway Usage Act a 30 year exemption from EU competition rules. In October 1996, the European Court annulled this decision. Following the Court's decision, the EC is reconsidering its original decision and may issue a new ruling. Thus may have the effect of reducing the proportion of capacity of the tunnel which is available for Le Shuttle services.
General economic risk	Revenue projections are determined in part by underlying economic activity in the UK and France and its attendant impact on business and leisure travel.
	The Group generates revenues and incurs costs in both French francs and sterling, although in practice the bulk of revenues will be generated in sterling, while the bulk of operating expenditure should be incurred in francs. This means that Group's operating results may be affected, positively or negatively, by fluctuations in exchange rates depending upon the Group's mix of revenues and costs at the time.
	Project economics also appear sensitive to changes in the rate of inflation and interest rates, which in the long run may be detrimental to Group economics.

new vehicles from the United Kingdom to Italy, while vehicles are transported to England from France, Italy and Belgium. Other services, retail operations, telecommunications and property development, currently account for only a modest share of total revenue.[19]

The first panel of Table 7.5 compares current revenue forecasts classified by Eurotunnel's component services with the project's original revenue estimates: in every case, the latest forecasts are only half of those of the initial forecasts. These errors appear due in the main to extremely optimistic assessments of freight usage; forecasts of passenger volumes, by contrast, are remarkably close to the original estimates, indeed they now favour higher usage than assumed initially. Freight projections, on the other hand, are extremely wide of the mark, with current forecasts looking for freight tonnages effectively one half of those originally estimated. This appears due in large part to stronger than expected competition from ferry operators, which both adjusted rates to minimise loss of market share and through rationalisation with its attendant favourable impact on operating cost structures.

Note, too, that market share data have been revised positively. Initial projections appear fairly conservative in respect of both passenger and freight volumes, which were expected to grow no faster than the total cross channel market. On the basis of current projections, both Eurostar and Le Shuttle are expected to increase market share up to the end of the decade.

The medium-term projections upon which these forecasts are based are shown in Table 7.6: current assumptions, with one exception, equal or exceed those used to generate the original base case. The sole significant exception is the exchange rate, which assumed an ongoing French franc–sterling rate of 8.5:1, more or less in line with the rate that emerged immediately following the collapse of the European Monetary System in mid-1992 rather than with the current rate. Otherwise, UK growth rates have been increased marginally, forecast inflation rates lowered – significantly from about 6 per cent per annum to around half that rate – as have interest rates. For the latest projections, growth rates, inflation and interest rates are assumed to be uniformly higher in the United Kingdom than France; over the longer term, the forecasts tend to converge. The second panel of the table relates longer-term economic projections into Eurotunnel operating results, and again the latest data, whether measured on a conservative or aggressive basis, are uniformly more optimistic than are the original projections.

In volume terms, forecast passenger growth rates are almost twice initial estimates, while the multiples applying to freight forecasts are higher still. Market shares, on the other hand, are not all that different, with Le Shuttle, regardless of the assumptions used, making deep inroads into both the passenger and freight segments; Eurostar, on the other hand, looks like being able just to hold on to market share, with the difference across the two scenarios shown of the order of roughly 2 percentage points. In financial terms, the data suggest total revenues ought to increase considerably more rapidly than originally assumed – 4 per cent versus just over 2 per cent; shuttle revenue growth is forecast to be well above original projections in comparison with freight forecast, which on the optimistic scenario is marginally higher than originally forecast but actually declines on the more pessimistic alternative. Finally, Table 7.7 presents ongoing risks to the project's future well-being.

The key point of the preceding analysis was to demonstrate how pervasive project risks can be even when revised projections of key operating and financial variables are uniformly more favourable than those used to generate the base case. In the case to hand, it was not so much a failure to predict traffic volumes, if anything the preceding analysis suggests original estimates were too conservative. Rather, the problem lies in being able to translate these results into accurate revenue and cash flow estimates. The main source of error was project capital cost: the actual outturn, at twice the originally estimated cost, imposed an enormous financial penalty on the project that sponsors have yet to sort out, notwithstanding a substantial refinancing that was intended to create the space needed to accomplish precisely that result.

PIPELINE PROJECTS

The pipeline industry was among the first of the regulated utilities to be liberalised. Deregulation first and foremost resulted in the creation of a competitive environment, mainly by altering the status and role of transmission companies within the industry. In the United States, for example, gas transmission companies were simultaneously monopsonists, being the only direct purchasers of natural gas in the economy, and monopolists, in that they were the sole sellers of gas to local distribution companies and other large industrial users. The prices received by pipelines for the gas shipped depended uniquely upon each company's cost of service, while purchasers were prohibited from sourcing supplies directly in field markets. Moreover, economic efficiency was severely impaired owing to the extremely complicated system of wellhead price controls then in effect.

Beginning in 1984, the Federal Regulatory Energy Authority (FERC) eliminated all wellhead price controls and conceded to local distribution companies and other industrial users the right to negotiate directly with gas producers to secure part or all of their natural gas requirements. Pipeline companies were now required by law to transport such quantities, with access to existing pipeline capacity conceded on a first come first service basis. The objective here was to ensure that those seeking transportation services only were not discriminated against since there was the reasonable assumption pipeline companies would favour buyers that continued to procure their gas supplies through traditional means.[20]

Natural gas markets have also been liberalised in Canada and the United Kingdom; the European Union, too, recently decided to deregulate the internal European market. In so far as we can tell, in not a single instance have regulatory authorities ever considered, except perhaps tangentially, the impact liberalisation might have had on the future development of transmission capacity. We saw in Chapter 5 the direst predictions of the consequences expected from decontrol, namely, that financial markets would categorically refuse to finance pipeline projects without secure long-term sales commitments. These were wrong, pure and simple.

There are, however, other effects, both obvious and subtle, that can be identified. The first, and perhaps most significant, relates to the nature of regulation itself. The behaviour of market participants (producers, transmission companies, distributors and purchasers of natural gas) is conditioned by the assumption that the prevailing market structure will remain more or less unchanged; in the event that changes do occur, there must be the reasonable expectation that such changes will be neutral in terms of their impact across all market segments. And second is the extent to which regulatory authorities are prepared to consider the importance of financial issues in the formulation of policy. The fact of a pipeline project having obtained financing does not, as some commentators assert, constitute proof that current regulatory policies successfully (or, rather inconsistently, have no need to) address financial matters.

We might note in this connection that the creditworthiness of shippers is one of the more important considerations taken into account by lenders in providing limited recourse finance for pipeline projects (Pollio, 1991a). If one of the original shippers assigns its capacity to a third party, lenders must be reasonably confident that the same shipper will retain its financial liability over the life of the financing. The critical point here is whether such continuing obligations are contingent upon the abandonment being temporary or permanent. The nature of the obligation is clear in respect of a temporary abandonment. On the other hand, it is difficult to imagine that a shipper, having filed for and obtained a permanent abandonment, would be willing to remain liable for its original financial commitment. This conflict raises several interesting questions. For an approved permanent abandonment, lenders would be willing to consider as substitute a shipper of equal or greater credit standing than the original shipper, otherwise the risk characteristics of the financing would change. On the other hand, the imposition of financial tests could be construed as discriminatory, a patent violation of the principle of open access that governs virtually all pipeline regulatory regimes.

The key point is that, as with infrastructure investments generally, the nature and consistency of regulatory policy is a leading, perhaps the key, risk faced by lenders. The ideal regulatory regime is one in which projects can count on the application of consistent and stable tariff payments. Lenders are also keen to ensure that regulatory authorities will allow for demand charges that cover a project's fixed costs, including all financial charges, and that any changes in the price of the commodity being shipped can be passed through to the ultimate purchasers. The importance of this point follows from the fact that some transmission companies, in addition to providing transportation, as was true in the United States, have a merchant function as well, so that failure to authorise full pass through could impair the pipeline's ability to meet debt service payments.

Apart from regulatory risks, throughput contracts are another critical risk factor affecting lender willingness to provide limited recourse project debt. There are two types of throughput contracts, each providing lenders with a different degree of comfort in respect of the financing. Firm capacity provides stronger financial assurances to creditors than does interruptible transportation. In the first instance, the buyer has contractually reserved pipeline capacity and in most instances will be under obligation to pay reservation charges whether the commodity is shipped or not. For interruptible supplies, the shipper arranges with the pipeline company the right to transport a commodity subject to capacity availability, with the shipping charge a function of prevailing demand for transportation services. For the first case, cash flow is only partially a function of the prevailing level of throughput, with lenders having additional financial cushion in the form of fixed reservation (or capacity) payments that are an integral part of the tariff structure. Otherwise, project revenue is a function of transportation only. Finally, lenders generally regard pipeline projects where sponsors have a direct interest in the facility as being financially superior to those that rely upon third party shippers; as we saw above, the creditworthiness of shippers is of prime importance to creditors, and connects with regulatory risks.

The competitive position of the project in the market which it serves is also of importance to lenders. The more transporters, actual or potential entrants, there are the riskier will be project cash flow as increased competition implies progressive downward pressures on the level of tariffs the project can charge. In the same vein, the type of commodity being shipped is of considerable importance: the greater is substitution potential – as would be the case with a natural gas pipeline if end users were able to switch to using fuel oil – the more volatile will project revenues be. Similarly, the comparative economics of alternative delivery transportation systems are bound to be of importance to end users. Thus projects where transportation charges are a significant component of total delivered cost are likely to be viewed less favourably by creditors than those where the opposite applies, on the not unreasonable assumption that such costs could undermine the competitive position of the shipper or shippers. For similar reasons, dedicated pipelines have a lower risk profile than those that are not, a conclusion that applies equally to projects able to interconnect with other pipeline systems.

In respect of capital and operating costs, pipeline projects are not all that different from other sectors we have looked at. Lenders prefer technically simpler to more complex projects and will virtually never give value to unproved technologies that are critical to meeting cost targets. Creditors are similarly keen that the project should be completed within the given construction period and at or below original capital cost estimates. The greater the risk such failures will be caused by, *inter alia*, delays in obtaining needed building permits or environmental compliance certificates, the stronger is the likelihood lenders will be inclined to view such delays as an event of default under the loan agreement.[21]

To illustrate the significance of the some of the points made above, we shall review the transportation and sales contract of the Transgas de Occidente Company (Colombia), both as means of showing how various contractual risks have been addressed and how the sponsors have dealt with currency risk (Simonson, 1997a). The project involves construction of a 344 kilometre mainline natural gas pipeline and is owned by TransCanada (34 per cent stake) and others. The project will after 20 years be transferred to Ecopetrol, the national oil company, for a payment equal to 1 per cent of project construction costs. Project cash flow is underwritten by a transportation service contract that requires Ecopetrol to pay upon project completion a monthly tariff whether the amount of gas designated for shipment by Ecopetrol is transported or not. On the other hand, the payment can fall to zero in the event the pipeline is not operational.

The transportation service contract is designed to ensure the project is able to cover all financial and operating costs, including the sponsor's required equity return. The tariff structure takes account of project availability, the monthly payment being the product of the agreed monthly tariff (initially set at $3.64 million) and project operating performance. The base tariff is predetermined and is the amount Ecopetrol has to pay Transgas. In practice, the monthly payment comprises both dollar and peso components, with each indexed to the relevant inflation index and to the exchange rate *vis-à-vis* the US dollar. The dollar component relates to debt service and other costs denominated in hard currency, while the peso rate is applicable to operating and maintenance costs incurred in local currency. The performance factor is defined as the ratio of the amount of gas delivered to the amount of gas scheduled to be delivered by Ecopetrol and may never exceed unity. Dollar costs are paid in local currency, with Ecopetrol assuming exchange rate risk providing Transgas has correctly managed and reported the resulting exposure.

Three main risks have been identified as being of critical importance to lenders.[22] The first relates to operating and maintenance expenditures, which under the transportation sales agreement are fixed. Thus any discrepancy between budgeted and actual expenses, other things being equal, would have a negative impact on project cash flow and by extension debt service payments. More positively, operating and maintenance expenses constitute a relatively small fraction (15 per cent) of the total tariff, meaning the project could tolerate a fairly wide variance in such costs before materially affecting financial results. Project cash flow is also at risk at capacity factors below 75 per cent, a wide margin of error given that pipelines are usually expected to be available virtually 100 per cent of the time.

The factors that might affect availability and how the costs if any are to be allocated between the project and Ecopetrol are far more critical. In the event that pipeline disruption is caused by earthquakes or so-called acts of God, Ecopetrol is not obligated to pay the monthly tariff, nor will it be liable for any repair costs. On the other hand, the throughput agreement upon termination of the original transportation and sales agreement can be extended for a length of time equal to the period the project was non-operational. If the disruption results in only partial suspension of pipeline operation, Ecopetrol is obligated to pay a portion of the monthly payment, the percentage equal to the fraction of service unaffected by the *force majeure* event, and extend the contract by the number of days service was affected. Finally, the base tariff is subject to renegotiation but only in response to changes in tax or environmental laws; all other contingencies, including significantly, a change in the law unrelated to tax or environmental matters, cannot trigger an increase in the monthly tariff even if such changes impair debt service.[23]

Notes

1   Infrastructure projects are normally developed as concessions. A concession is effectively a licence, usually granted by a government or government agency, that gives the concessionaire the right to build the relevant infrastructure and operate it for a pre-determined period at the end of which the project is transferred to the government issuing the concession (Wood and Vintner, 1992). This structure, however, is not invariant: the concession to construct the Shajiao B Power plant was granted via a Chinese cooperative joint venture contract between the responsible local enterprise and the foreign investment vehicle (Clifford Chance, 1991).
2   The Sutton Bridge power financing is an interesting transitional example in that post-2014, if the existing Capacity and Tolling Agreement is extended, creditors will be exposed to merchant risk over the remaining term of project debt (Rigby, Simonson and Wilkins, 1997).
3   As we saw in Chapter 3, international capital budgeting theory limits project cash flows to those that can be remitted freely. Loss of convertibility could, at a stroke, wipe out a large fraction of project cash flow. Within a limited recourse debt framework, that would be sufficient reason for sponsors to default.

4 Let P_{ct} and P^*_t be, respectively, the local currency and US dollar price of a Big Mac in year t, and X_t is the spot exchange rate, defined as the local currency cost of one US dollar. Big Mac purchasing power parity is defined as: $r_{ct} = \ln(P_{ct}/P^*_t/X_{ct}) = \ln(P_{ct}/P^*_t) - \ln X_{ct}$. If $r_{ct} > 0$, then the local currency is said to be overvalued relative to the US dollar; if, by contrast, $r_{ct} < 0$, then the currency would be considered undervalued. Only where $r_{ct} = 0$, that is, where $\log(P_{ct}/P^*_t) = \log X_{ct}$ does purchasing power parity apply.

5 The forward price parity involves two interest rates, r_H and r_F, the interest rate applying in the investor's home market and the rate in the target foreign market, respectively and the current spots price, S. The forward price parity can be expressed as $F/(1 + r_H) = S/(1 + r_F)$ or $F = S(1 + r_F)/(1 + r_H)$. In words, the forward rate (measured in terms of the number of foreign units per unit of home currency) will be at a discount *vis-à-vis* the current spot rate and by an amount equal to the prevailing interest differential.

6 The expectations hypothesis requires that $F = E(S)$, where the variables are as before and E is the expectations operator. If we substitute $S(1 + r_F)/(1 + r_H)$ for $E(S)$, then it follows that the best prediction of future spot rates will be either the forward rate or the ratio of domestic to foreign interest rates.

7 The ability of the international financial community to manage the crisis depends less upon the financial standing of the vehicle through which support is provided, usually the International Monetary Fund, than upon the magnitude of the financial commitment that leading OECD countries area prepared to make.

8 Blaiklock (1998) argues that recent attempts by a number of host countries to shield infrastructure projects from currency risk by agreeing to the creation of standby facility intended to cover foreign exchange shortfalls appears counter-productive. The main concern here is that the creation of what amounts to second layer of subordinated debt provided by the host government is a hidden subsidy or, alternatively, a means of converting direct financial obligations into contingent liabilities. In extreme cases, standby facilities can give the appearance of financial viability to inherently weak projects.

9 Daily price fluctuations in the US power market can be as high as 300 per cent (Rigby, 1997b).

10 However beneficial in competitive terms new generating technologies may be, such considerations are unlikely to encourage lenders to alter existing credit principles that discount unproved technology, especially where such technology is crucial to achieving specific revenue or cost targets.

11 Safety considerations are often given to justify the exclusion of private operation and control of air traffic assets. Such justifications are hollow in that there is no evidence to suggest private operational control is any less efficient than some form of public sector management. The sacking by the Reagan Administration of air traffic controllers and their replacement by non-public sector employees illustrates the point. In fact, the opposite is more nearly true: private sector management normally brings both greater financial and operational efficiency.

12 CODAD is owned 66 per cent by Dragados (Madrid, Spain), 19 per cent by Ogden Corporation (US) and 15 per cent by Conconcreto Ingenieros Civiles SA (Colombia). No transfer of ownership is permitted prior to completion of the second runway; post-completion, the consortium must maintain management control of CODAD and either Dragados or Ogden must individually maintain a 25 per cent stake. The project is capitalised at \$145 million, with equity of \$29 million (20 per cent) and \$116 million in debt in the form of senior notes; three quarters of the total covers the fixed price EPC contract with capitalised interest accounting for another 14 per cent.

13   The flow of funds generated by airline payments is allocated first to cover the following month's expected operation and maintenance expenses. Any remaining amounts are dispatched to an offshore trustee which divides these receipts between the debt payment account, the debt service reserve account (funded upfront to provide 6 months debt service cover), and major maintenance account. Any residual sums can then be distributed to CODAD.

14   A prime example of the deleterious impact current work rules can have on port efficiency is the fact that container handling costs at Alexandria, Egypt's main port, are roughly two to three times higher than at other Mediterranean ports. These extra costs are obviously a deadweight cost to the economy (Pollio, 1996).

15   Actual usage on the Hardy Toll Road has been about 50 per cent below original traffic volume forecasts. Other examples pointing in the same direction are the Bangkok Tollroad and Mexican toll roads. Blaiklock (1998) maintains toll road liabilities were one of the main drivers of the Mexican financial crisis in the early 1990s.

16   Toll rates are assumed to rise in keeping with inflation and other factors that have a bearing on operating and maintenace costs. Thus base case economics assume no real increase in tolls over the life of the concession.

17   Another unfortunate feature of the way base case economics were assessed is that growth rates are assumed to decline progressively from current rates over the life of the concession. For the three regions covered by these projects economic growth rates are forecast to decelerate across the board from the actual 10 per cent rate achieved over 1989–95 period to around 4 per cent or less for the 2020–30 period. Even on what are considered to be 'conservative' forecasts, the differences amount to around 0.5 per cent per annum over concession life (Doud and Forsgren, 1997). As road usage is a function of regional and national economic growth rates, the financial impact of the assumptions used to generate the base case is to front load traffic volumes. This of course translates to a higher present value of potential project revenue. The current economic outlook would seem to imply an inverted 'V' profile, with growth (and by extension, traffic volumes) higher in the outyears than over the near and medium term. Combined with the risk that the government is less likely to authorise compensatory toll increases, project cash flows look to be at even greater risk than the worst case scenario described above.

18   In the case of the United Kingdom, track and signalling services are provided by Railtrack. Concessions involving the London Underground are even more complex, all the more surprising in that coordination of the various aspects of local transportation policy are to be vested in the soon to be created post of mayor. Could it be the government is keen to have the postholder assume responsibility for what is rightly seen as being an extremely confused policy *vis-à-vis* London's subway system?

19   Retail concessions pay Eurotunnel a rent calculated on either a fixed or percentage of sales basis. Other sources of revenue consist of the sale of own brand merchandise, fees collected from vendors selling fuel or other goods and fees for the use of space in the terminals or on board for advertising purposes. Property development is concentrated on the 207 hectares of land managed or controlled by Eurotunnel. Development is to be based upon the sale or lease of land or buildings where the proposed activities either upgrade areas adjacent to the terminals or enhance prospective tunnel use.

20   'When the change in policy had the desirable and intended effect of reducing the price of gas below the inflated level created by the prior policy, producers and pipelines had to absorb scores of billions of dollars in the transition from one

policy to the other. This produced financial hardship, thousands of contract disputes, and major political and legal controversies' (Pierce, 1993). Many of these problems could have been avoided. FERC had the statutory authority either to suspend the contracts creating the take-or-pay obligations or, alternatively, impose a one time surcharge on natural gas purchases equal to the outstanding take-or-pay liabilities. In the event, the FERC opted to do nothing, instead encouraging producers and pipeline companies to renegotiate existing contracts. Hence the importance of a more even-handed approach to market decontrol.

21   Among other factors worth noting, mention should be made of the importance of the climate or terrain over which the pipeline runs, affecting as it does operating and maintenance expense, utilisation and the physical condition of the pipe; insurance to deal with various contingencies unique to pipeline investments; the extent to which the project might be expected to match or exceed that of other similar pipeline investments; and the reputation of the project operator (Simonson, 1997, who also presents a full discussion of financing criteria for pipelines from the perspective of an international credit rating agency).

22   Debt service risks were addressed in what would appear to be a less than straightforward manner. The financing agreement provides for the creation of a fully funded 1 year debt service reserve which declines to 6 months upon 12 months successful project operation. Under normal circumstances it would not be unreasonable to permit an eventual lowering of the reserve since the principal risks affecting debt service should be known 1 year after project commencement. In the case to hand, the low after-tax debt service coverage ratios, below $1.5\times$ in the base case and $1.36\times$ in the low scenario, would seem to argue in favour of the desirability of maintaining the larger debt service reserve for longer.

23   'Change in the law risk... must be adequately covered to reach investment-grade credit strength. Change in the law risk was mitigated by the acceptance of this risk on a joint and several basis by the two strongest rated sponsors of the... project in the event a change in the law reduces debt service coverage below adequate levels' (Simonson, 1997a).

References

Abel and Blanchard (1986) 'The Present Value of Profits and Cyclical Movements in Investment', *Econometrica*, **54**.

Adelman (1993) 'Modeling World Oil Supply', *Energy Journal*.

Adewole (1992) 'OPEC Natural Gas Exports: Past, Present and Future', *OPEC Review*, **16**.

Agmon and Lessard (1977) 'Investor Recognition of Corporate International Diversification', *Journal of Finance*, **32**.

Alexander (1995) *Cost of Capital* (Oxford Economic Research Associates Ltd).

Allen (1996) '$570m Loan for Qatar Gas Project', *Financial Times*, 23 December.

Altman (1984) 'A Further Empirical Investigation of the Bankruptcy Cost Question', *Journal of Finance*.

Altman and Subrahmanyam (1985) *Recent Advances in Corporate Finance* (Richard D. Irwin).

Alzard (1996) 'Technical and Scientific Progress in Petroleum Exploration-Production: Impact on Reserves on Costs', *Energy Exploration and Exploitation*.

Anderson and Gilbert (1988) 'Commodity Agreements and Commodity Markets', *Economic Journal*, **98**.

Bekaert and Harvey (1995) 'Time-Varying World Market Integration', *Journal of Finance*, **50**.

Barro (1990) 'The Stock Market and Investment', *Review of Financial Studies*, **3**.

Baxter (1996) 'European Refinery Investment: A Bankers Perspective', Unpublished, Chase Manhattan Bank, November.

Bernstein (1996) *Against the Gods* (Wiley).

Black (1993) 'Beta and Return', *Journal of Portfolio Management*, Fall.

Blaiklock (1993) Financing Infrastructure Projects as Concessions. PIRC Annual Conference.

Blaiklock (1998) Financing Transportation Infrastructure Projects as Investments, unpublished.

Blanchard (1997) *Macroeconomics* (Prentice Hall International).

Bower *et al.* (1998) 'A Primer on Arbitrage Pricing Theory', in Stern and Chew (eds) *The Revolution in Corporate Finance* (Blackwell).

Brealey, Cooper and Habib (1996) 'Using Project Finance to Fund Infrastructure Investments', *Journal of Applied Corporate Finance*, **9**.

Brealey and Myers (1991) *Principles of Corporate Finance*, 4th edn (McGraw Hill).

British Petroleum (1997) *Statistical Review of World Energy*, June.

Brown (1990) *OPEC and the World Energy Market*, 2nd edn (Longman).

Buckley (1996a) *Multinational Finance*, 3rd edn (Prentice Hall).

Buckley (1996b) *International Capital Budgeting* (Prentice Hall).

Burgman (1996) 'An Empirical Examination of Multinational Corporate Capital Structure', *Journal of International Business*, **27**.

Butters, Fruhan, Mullins and Piper (1987) *Case Problems in Finance*, 9th edn (Irwin).

Castle (1986) 'North Sea Score Card', SPEE 15358, Society of Petroleum Engineers (October).

CBI (1994) 'Realistic Returns: How Do Manufacturers Assess New Investment?', Confederation of British Industries.

Chan and Lakonishok (1993) 'Are the Reports of Beta's Death Premature', *Journal of Portfolio Management*, Summer.

Chapman (1995) 'The Principles of Project Finance', *The Treasurer*, November.

Chemmanur and John (1992) 'Optimal Incorporation Structure and Debt Contracts and Limited Recourse Project Financing', New York University Working Paper.

Chen, Cheng, He and Kim (1997) 'An Investigation of the Relationship Between International Activities and Capital Structure', *Journal of International Business Studies*, **28**.

Chen, Roll and Ross (1986) 'Economic Forces and the Stock Market', *Journal of Business*, July.

Clifford Chance (1991) *Project Finance* (February).

Cohen (1992) *Athenian Economy and Society: A Banking Perspective* (Princeton University Press).

Commonwealth Development Corporation (1996) 'FFC-Jordan Fertiliser Co, Ltd', May.

Cox (1983) 'Financing an Oil Company', in McKechnie (1983a).

Creath (1983) 'Patterns in US Petroleum Lending', in McKechnie (1983a).

Dimson (1989) 'The Discount Rate for a Power Station', *Energy Economics*, July.

Dixit and Pindyck (1994) *Investment Under Uncertainty* (Princeton University Press).

Dixit and Pindyck (1995) 'The Options Approach to Capital Investment', *Harvard Business Review*, May–June.

Dols and Page (1998) 'New Markets Pose Different Challenges', *Petroleum Economist*, April.

Doud and Forsgren (1997) 'Greater Bejing First Expressways Ltd.', in Standard & Poor (1997).

Drew (1998) 'Latest Financing Techniques for Infrastructure Projects in Asia' (unpublished HSBC), March.

Emery and Finnerty (1997*) Corporate Financial Management* (Prentice Hall).

Eurotunnel plc (1997) Financial Restructuring Proposals, May.

Fama and French (1992) 'The Cross-Section of Expected Stock Returns', *Journal of Finance*, **47**.

Figlewski (1990) 'Basic Price Relationships and Basic Trading Strategies', in Figlewski, Silber and Subrahmanyam (eds) *Financial Options: From Theory to Practice* (New York University Press).

Finnerty (1996) *Project Financing* (John Wiley).

Fitzgerald and Pollio (1982) 'Aluminum: The Next Twenty Five Years', *Journal of Metals*, December.

Fitzgerald and Pollio (1984) 'Financing the Next Generation of Copper Projects*', Natural Resources Forum*, October.

Gas de France (1997) 'GNL: Horizons Lointains', *Gas du Monde*, September.

Gewirtz (1995) 'Legal Aspects of Project Finance', *The Treasurer*, November.

Godfrey and Espinosa (1996) 'A Practical Approach to Calculating Costs of Equity for Investments in Emerging Markets', *Journal of Applied Corporate Finance*, **9**.

Gotaas-Larsen (1997) *LNG World Overview*.

Groenwold and Fraser (1997) 'Share Prices and Macroeconomic Factors', *Journal of Business Finance and Accounting*, **24**.

Gwartney, Lawson and Black (1996) *Economic Freedom of the World* (Fraser Institute).

Haque, Kumar, Marx, and Mathieson (1996) 'The Economic Content of Indicators of Developing Country Creditworthiness', IMF *Staff Papers* 43.

Harries (1990) 'The Contract Law of Project Financing'.

Hayes and Garvin (1982) 'Managing as if Tomorrow Mattered', *Harvard Business Review*, **50**.

Higson (1991) *Business Finance* (Butterworth).

Howell and Jagle (1997) 'Laboratory Evidence on How Managers Intuitively Value Real Growth Options', *Journal of Business Finance and Accounting*, **8**.

HSBC (1997) *Capital Markets Funding for Limited Recourse Project Debt: A Guide to the Rule 144A Market*, Hong Kong Shanghai Bank.

Inselbag and Kaufold (1997) 'Two DCF Approaches for Valuing Companies Under Alternative Financing Strategies (And How to Choose Between Them)', *Journal of Applied Corporate Finance*, **10**.

International Energy Agency (1991) *Natural Gas: Prospects and Policies* (OECD).

Jensen and Meckling (1976) 'Theory of the Firm: Managerial Behavior, Agency Costs and Ownership Structure', *Journal of Financial Economics*, **11**.

John and John (1991) 'Optimality of Project Financing: Theory and Empirical Implications in Finance and Accounting', *Review of Quantitative Finance and Accounting*, **1**.

Kemna (1993) 'Case Studies on Real Options', *Financial Management*, Autumn.

Kim and Singal (1997) 'Are Open Markets Good for Foreign Investors and Emerging Nations?', *Journal of Applied Corporate Finance*, **10**.

Kleimeier and Megginson (1996) 'An Economic Analysis of Project Finance', Paper presented at the City University Business School, March.

Kleimeier and Megginson (1997) 'A Comparison of Project Finance in Asia and the West', in Lang (ed.) *Project Financing in Asia: A Redefining of Premises* (North Holland).

Koiishi (1983) 'Financing the Indonesian LNG Developments', in McKechnie (1983a).

Kopcke (1993) 'The Determinants of Business Investment: Has Capital Spending Been Surprisingly Low?', *New England Economic Review*, January/February.

Krugman (1998) 'What Happened in Asia?', cited in the *Economist*, January 10.

Kulatilaka and Marcus (1992) 'Project Valuation Under Uncertainty: When Does DCF Fail?', *Journal of Applied Corporate Finance*, **5**.

Lawden (1968) 'Modeling Physical Reality', *Philosophical Journal*, **5**.

Leslie and Michaels (1997) 'The Real Power of Real Options', *McKinsey Quarterly*, **3**.

Lessard (1996) 'Incorporating Country Risk in the Valuation of Offshore Projects', *Journal of Applied Corporate Finance*, **9**.

Lessard and Shapiro (1983) 'Guidelines for Global Financing Choices', *Midland Corporate Finance Journal*, **1**.

Malkiel and Xu (1997) 'Risk and Return Revisited', *Journal of Portfolio Management*, Spring.

Marsh (1998) 'Power Equipment Orders From SE Asia Set to Fall Sharply', *Financial Times*, 4 June.

Mason and Merton (1985) 'The Role of Contingent Claims Analysis in Corporate Finance', in Altman and Subrahmanyam (1985).

McKechnie (ed.) (1983a) *Energy Finance* (Euromoney).

McKechnie (1983b) 'Oil and Gas Field Development Finance', in McKechnie (1983a).

McKechnie (1990) 'Project Finance', in Terry (1990).

McTiernan (1983) 'Reserves and Production Risk in Petroleum Financing', in McKechnie (1983a).

Merrill Lynch (1997) *Global Research Review*, June.

Middlemann (1996) 'Bond Financing Catches On', *Financial Times*, 3 December.

Mikesell and Whitney (1987) *The World Mining Industry: Investment Strategy and Public Policy* (Allen & Unwin).

Modigliani and Miller (1958) 'The Cost of Capital, Corporation Finance and the Theory of Investment', *American Economic Review*, **48**.

Modigliani and Miller (1963) 'Corporate Income Taxes and the Cost of Capital: A Correction', *American Economic Review*, **53**.

Mueller (1985) 'Scarcity and Ricardian Rents for Crude Oil', *Economic Inquiry*, **23**.

Myers (1974) 'Interactions of Corporate Financing and Investment Decisions – Implications for Capital Budgeting', *Journal of Finance*, **29**.

de Nahlik (1992) 'Project Finance', in Rutterford and Montgomerie (eds) *Handbook of UK Corporate Finance,* 2nd edn (Butterworth).

Odell and Rosing (1983) *The Future of Oil* (Kogan Page).

Ong (1997) 'Burgernomics: The Economics of the Big Mac Standard', *Journal of International Money and Finance*, **16**.

Paddock, Siegel and Smith (1988) 'Option Valuation of Claims on Real Assets: The Case of Offshore Petroleum Leases', *Quarterly Journal of Economics*, **103**.

Petroleum Economist (1998) *The Fundamentals of the Global LNG Industry.*

Petroleum Review (1998) 'South Korea Reviews Gas Growth Forecast', April.

Pierce (1993) Experiences with Natural Gas Regulation and Competition in the US Federal System: Lessons for Europe', in Mestmacker (ed.) *Natural Gas in the Internal Market: A Review of Energy Policy* (Graham and Trotman).

Pindyck (1991) 'Irreversibility, Uncertainty and Investment', *Journal of Economic Literature*, **29**.

Pollio (1985) *Empirical Tests of the Rational Expectations Hypothesis*, PhD dissertation, City University Business School.

Pollio (1986) 'Financing Energy Development: Trends and Prospects', *Journal of Energy and Development,* **12**.

Pollio (1990) 'The Application of Commodity Linked Products to Project Financings', Paper presented at the *Euromoney* Conference on Financing and Hedging with Commodity Linked Products, London, February.

Pollio (1991) *Financial Innovation and Upstream Petroleum Development*. International Research Centre for Energy and Economic Development, University of Colorado, Occasional Paper Number 14.

Pollio (1991a) *The US Natural Gas Market in the 1990s*, International Research Center for Energy and Economic Development, University of Colorado, Occasional Paper Number 12.

Pollio (1992) 'Constraints on the Oil Industry in the 1990s: The Financial Dimension', *OPEC Review*, **16**.

Pollio (1993) 'Financing Energy Development in the 1990s', *Natural Resources Forum*, August.

Pollio (1994) 'Qatargas Downstream Venture: Structure and Financial Analysis', Qatar General Petroleum Corporation, June.

Pollio (1996) 'Restructuring in an Emerging Market: A Sectoral Analysis of the Egyptian Economy', City University Business School, unpublished, July.

Pollio and Hart (1994) 'North Field Condensate Enhancement Options', Qatar General Petroleum Corporation, August.

Poten and Partners (1991) *New LNG Trades: Costs and Competition.*

Power in Asia (1998) 'South Asia Throws a Wobbly', *Financial Times, Power In Asia*, 18 May.

Priovolos and Duncan (1991) *Commodity Risk Management and Finance* (Oxford University Press).

Project Finance (1998) *Project Finance*, January.

Rigby (1997) 'Currency-Risk Management: Key Criteria', in Standard & Poor (1997).

Rigby (1997a) 'Project Financed Transactions: Technical Risk Assessment', in Standard & Poor (1997).

Rigby (1997b) 'Merchant Power Plants: Project Finance Debt Criteria', in Standard & Poor (1997).

Rigby, Magtoto, Penrose and Fukutomi (1997) 'Ras Laffan Liquified Natural Gas Company Ltd.', Standard & Poor's *Global Project Finance*, March.

Rigby, Simonson and Wilkins (1997) 'Sutton Bridge Financing Limited', in Standard & Poor (1997).

Rose (1992) 'Project Finance: The Way Forward in Eastern Europe', *Petroleum Economist*, International Energy Law Supplement, September.

Rosenberg and Rudd (1998) 'The Corporate Uses of Beta', in Stern and Chen (eds) *The Revolution in Corporate Finance*, 3rd edn (Blackwell).

Sayer (1997) 'A Storm of Activity', *Trade and Project Finance*, March.

Scarlett (1990) 'International Capital Markets – Bonds', in Terry (1990).

Shah and Thakor (1987) 'Optimal Capital Structure and Project Financing', *Journal of Economic Theory*, **42**.

Shapiro (1989) *Multinational Financial Management*, 2nd edn (Allyn & Bacon).

Shell (1992) 'World LNG', *Shell Briefing Service*, **4**.

Siegel, Smith and Paddock (1987) 'Valuing Offshore Oil Properties with Options Pricing Models', *Midland Journal of Corporate Finance*, **5**.

Simonson (1997) 'Criteria for Project Financing of Pipelines', in Standard & Poor (1997).

Simonson (1997a) 'Transgas De Occidente', in Standard & Poor (1997).

Simonson and Haddad (1997) 'Compania De Dessarollo Aeropuerto Eldorado S.A.', in Standard & Poor (1997).

Smith (1986) 'Applications of Option Pricing Analysis', in Jensen and Smith (eds) *The Modern Theory of Corporate Finance* (McGraw-Hill).

Smith (1995) 'The Hub Power Project – The Anatomy of Sponsorship', *The Treasurer*, November.

Smith (1996) 'Bank of America Roundtable on Evaluating and Financing Foreign Direct Investment', *Journal of Applied Corporate Finance*, **9**.

Smith and Walter (1990) *Global Financial Services* (Harper).

Smith and Walter (1997) *Global Banking* (Oxford University Press).

Spence and Godier (1996) 'Powerful Sources of Support', *Financial Times*, 3 December.

Standard & Poor (1996) *Global Project Finance*, July.

Standard & Poor (1997) *Global Project Finance*, September.

Statman (1987) 'How Many Stocks Make a Diversified Portfolio?', *Journal of Financial and Quantitative Analysis*, **22**.

Stauffer (1996) *The Diseconomies of Long Haul LNG Trading*, International Research Center for Energy and Economic Development, University of Colorado, Occasional Paper Number 26.

Stern (1992) *Third Party Access in European Gas Industries*, Royal Institute of International Affairs.

Stollery (1983) 'Mineral Depletion with Cost as the Extraction Limit: A Model Applied to the Behavior of Prices in the Nickel Industry', *Journal of Environmental Economics and Management*, 10.

Streifel (1995) *Review and Outlook for the World Oil Market*, World Bank Discussion Paper Number 301.

Summers (1987) 'Investment Incentives and the Discounting of Depreciation Allowances', in Feldstein (ed.) *The Effects of Taxation on Capital Accumulation* (University of Chicago Press).

Symon (1997) 'Foundation for Regional Growth', *Financial Times*, International Gas Industry Survey, June 10.

Taylor (1996) 'Entrepreneurial Skills Needed', *Financial Times*, December 3.

Terry (ed.) (1990) *International Finance and Investment*, 2nd edn (Chartered Institute of Bankers)

Thackeray (1998) 'OPEC's Future Rests on Asian Tigers' Return to Growth', *Petroleum Review*, July.

Thomas (1998) 'Indonesia: Maintaining Its Position as World's Largest LNG Exporter', *Petroleum Economist*, April.

Trigeorgis (1993) 'Real Options and Interactions with Financial Flexibility', *Financial Management*, Autumn.

Tussing (1996) 'Gas Prices Will Fluctuate', *Energy Exploration and Exploitation*, **14**.

Van Horne and Wachowicz (1995) *Fundamentals of Financial Management*, 9th edn (Prentice Hall).

Verleger (1993) *Adjusting to Volatile Energy Prices* (Institute for International Economics).

Walde (1992) 'Recent Developments In Negotiating International Petroleum Agreements', *Petroleum Economist*, International Energy Law Supplement, September.

Walde (1993) 'Comment on Einer Hope's Chapter with a Critical Review of Legal and Policy Arguments Driving the Discussion on Third Party Access', in Mestmacker (ed.) *Natural Gas in the Internal Market* (Graham and Trotman).

Warner (1977) 'Bankruptcy Costs: Some Evidence', *Journal of Finance*, **32**.

Weston and Copeland (1998) *Managerial Finance* (Cassell).

Woicke (1983) 'LNG Financing', in McKechnie (1983a).

Wolf (1998) 'Uphill Struggle', *Financial Times*, 2 June.

Wood (1980) *Law and Practice of International Finance* (London: Sweet & Maxwell).

Wood and Vintner (1992) *Legal Aspects of Project Finance* (Allen & Overy).

Index

Page numbers in *italic* denote tables; those in **bold** refer to charts and figures. The letter n following a page number denotes the note number on that page.